P9-CFO-206

the tiger on your couch

the tiger on

your couch

WHAT THE
BIG CATS CAN TEACH YOU
ABOUT
LIVING IN HARMONY
WITH
YOUR HOUSE CAT

•

bill fleming
judy petersen-fleming

William Morrow and Company, Inc.
New York

Copyright © 1992 by Bill Fleming and
Judy Petersen-Fleming

Permissions and photograph credits,
constituting a continuation of the
copyright page, appear on pages
214 to 218.

All rights reserved. No part of this book
may be reproduced or utilized in any
form or by any means, electronic or
mechanical, including photocopying,
recording, or by any information storage
or retrieval system, without permission in
writing from the Publisher. Inquiries
should be addressed to Permissions
Department, William Morrow and
Company, Inc., 1350 Avenue of the
Americas, New York, N.Y. 10019.

Library of Congress
Cataloging-in-Publication Data

Fleming, Bill.
The tiger on your couch : what the big
cats can teach you about living in
harmony with your house cat / Bill
Fleming, Judy Petersen-Fleming.
p.   cm.
Includes index.
ISBN 0-688-08750-7
1. Cats—Behavior. 2. Cats.
3. Tigers—Behavior.
4. Tigers—Anecdotes. 5. Fleming, Bill.
I. Petersen-Fleming, Judy. II. Title.
SF446.5.F54 1992
636.8—dc20   91-32051
                                              CIP

Printed in the United States of America

Book Design by HARAKAWA SISCO INC.

To Baghdad,
for your wisdom

To Harriet,
for your patience

To Lisa,
for your inspiration

To Mark,
for your creativity

To Will,
for your encouragement

ACKNOWLEDGMENTS

A special thanks to all of the trainers on Tiger Island. Their
dedication to the cats and their efforts to educate people about the
big cats inspire the message of this book: that special bonding
between man and cat is always possible. Thank you Pat, Jim, Rick,
Jeff, Andy, and Gregg.

And a very special thanks to Marine World Africa USA for the
wonderful work they do and for generously giving permission to
use all of the tiger photos. All of these photos were taken at
Marine World Africa USA's Tiger Island.

contents

**NOTE TO THE READER
FROM JUDY**   10

INTRODUCTION   11

1 CLOSE COUSINS

A Whisker Away from the Wilds   15

Physical Similarities   16

Tiger Island   24

The Tiger on Your Couch   26

2 CHOOSING YOUR FELINE COMPANION

Does Your Life-style Fit a Cat and Does
a Cat Fit Your Life-style?   29

Are Two Cats Better Than One?   34

Choosing a Kitten or Cat   36

Healthy Signs   39

Getting the Pick of the Litter   44

Where to Find Your Feline   45

Tigers As Pets?   48

3 STARTING WITH THE BASICS

Remember the Predator   51

Kitten Conditioning: Taking Over the
Mother's Role   53

When Is the Best Time to Start
Conditioning Your Kitten?   56

ABCs of Conditioning   58

Get a Grip on Teeth and Claws   64

Why the *@%* Did He Do That?
Reasons for Unexpected Behavior   69

4 PLAYTIME WITH A FURRY FRIEND

Why Is Playtime Important?   73

Playtime Etiquette   75

Where to Play   78

Relieving Boredom   79

Terrific Toys   84

A Fit Cat   88

5 CAT NAPS

Why Does My Cat Take Sixteen-Hour Naps? 91

Sharing Quiet Times 94

Cat Talk 96

Building Closer Bonds with Touch 97

How to Make a Cozy Cat 98

"Leave Me Alone" Signals 101

Feline Physical During Quiet Times 102

Quiet Time and Grooming 104

Quiet Time and Fleas 105

The Benefits and Stress Releases *We* Gain from Sharing Quiet Time 107

Meditation with—and for—Your Cat 107

Partner Breathing with Your Cat 108

6 YOU AND YOUR INDOOR CAT

The Great Indoors 113

Cat Box and Spraying 114

Litter Box Sharing 117

Your Indoor Cat and Spraying 117

Household Hazards 120

The Couch Isn't a Scratching Post 121

That Was My Favorite Vase! 124

Plant Vandals 124

Feeding Facts 126

Feeding Advice 127

Feeding Transitions 129

Your Finicky Cat 129

Nourishment No's 130

Kitchen and Dinnertime Tips 130

Cat Calls 135

Cat Treats 138

7 CAT'S NIGHT OUT— YOUR INDEPENDENT INDOOR/OUTDOOR CAT

The Decision of an Indoor Versus Indoor/Outdoor Cat 141

How to Introduce a Cat to the Great Outdoors 143

In and Out and In and Out . . . 147

Let Them Know It's Suppertime! 151

Common Cat Fights: What to Do About Them 153

Rodents as Gifts 156

Birds 157

Cat on a Leash 158

Bathing: A Fun Day Out May Result in the Need of a Bath! 161

8 CAT ON THE MOVE

Should Cats Travel?   165

Travel Box Conditioning   166

Car Travel   170

Traveling by Air   173

Trip to the Veterinarian   174

9 YOUR CAT IN NEW SITUATIONS

Changing Routine   177

Moving   178

Introducing Your Cat to a New
Household Pet   180

Company Coming and New Household
Members   191

10 CAT CONSERVATION: WE CAN MAKE A DIFFERENCE

Cat Conservation   197

Starting in Your Own Backyard—
Spaying and Neutering   198

Environmentally Safe Pet Products
and Practices   200

The Endangerment of the Tiger   203

The Survival of Exotic Cats in America 206

A Tender Time with a Tiger   208

What to Do?   209

APPENDIX   210

PERMISSIONS AND CREDITS   214

INDEX   219

NOTE TO THE READER FROM JUDY

Bill and I are lifelong cat lovers—but only Bill has worked and lived
with the big cats. *The Tiger on Your Couch* is filled with stories
about the big cats that will show you that the difference between
your house cat and a tiger is largely a matter of size. Bill tells you
these stories himself, in his own words, separate from the rest of the
text. If you love tigers, you may also just want to go through the
book and read all the "first-person" stories about life with the wonderful tigers Bill has grown to know and love.

introduction

Tigers are very special animals: majestic, powerful, mysterious, affectionate, and funny. We consider ourselves extremely fortunate to have been able to spend so much time with them. Bill has watched and assisted (if needed) when they gave birth. We have observed them shift from the clumsiness of adolescence to the grace and dignity of adulthood. We have also seen the true playful side of a tiger as he chases his tail, and the irresistible look of a tiger asleep with her tongue sticking out. The domesticated cat does not have the dramatic size and power of his large ancestors, but he is every bit as majestic, mysterious, funny, and affectionate. When a cat strikes an artful pose of grace and beauty, or dashes in a mad frenzy in pursuit of imaginary prey, the effect is as hypnotic whether it is a house cat or tiger.

The goal of this book is to teach you how to have a happy and natural relationship with your house cat. Living harmoniously with a tiger and cat go paw in paw. Throughout the book we will be telling tiger tales—facts about big cat behavior as well as intimate details about the individual tigers Bill has worked with for more than a decade. These specific Bengal tigers that Bill has grown to know so well show the entire range of cat personality. You will be able to compare them with your house cat and learn from the similarities.

The ten tigers we will use for most of the examples range from ol' lady Baghdad, a seventeen-year-old female, to the youngster Sampson, a spunky three-year-old white Bengal tiger. Kimra is an aloof tiger, Tara a never-ending "talker," Mohan strong and noble, and Rakon a clumsy and lovable tiger. Through the years that Bill has been with these cats, he has witnessed every stage of a tiger's life— from the boundless energy of a cub to the awkwardness of adolescence, and then from maturity to the poised dignity of old age.

All cats big and small react to certain types of stimuli in a similar fashion. By understanding the behavior and antics of these ten tigers as well as of all tigers in the wild, you will better your understanding of your own cat's nature.

By learning to stimulate your cats, you can enjoy seeing the spark of true excitement in their eyes—and the contented look of exhaustion after an energetic play session. The more stimulation a cat receives, the longer, healthier, and more purposeful life she will lead. This book is not about the physical likenesses between tigers and cats; its purpose is to show you how similar the behavior of a cat is to a tiger's behavior. By understanding the wild instincts of a tiger, you will be able to understand the untamed and wild side that *all* domestic cats still possess.

There are staggering numbers of cats euthanatized each year because of abandonment. One of the primary reasons people abandon their cats is because of the cats' behavioral problems. People do not take the time to understand how to live comfortably with a cat.

We intend to concentrate on easy methods for you to use with your cat that will help you live together happily and in harmony. And

A BENGAL TIGER LIVES 10 TO 12 YEARS IN THE WILD AND AN AVERAGE OF 20 YEARS IN CAPTIVITY.

●

we will show you, step-by-step, foolproof ways to put them to use—
with a new kitten or longtime companion. We have learned through
the years that people and tigers can share their lives together with a
very special bond for one another. We are all able to share this same
bond with our cats at home by using exactly the same methods that
we have used to gain the tigers' trust. Your cat does not have to be
aloof and independent if you do not want him to be. By using these
simple techniques, and spending quality time with your cat, you can
begin to share a warm, affectionate, and dependable closeness with
her you may never have thought possible. And behavioral problems—like finicky eating or ruining the furniture—will be a thing of
the past.

   If you are a new cat owner, you will be able to use this book
as a guide for everything you need to know. Follow the techniques in
the chapters from start to finish. If you have always been a cat owner
and want to know how to have a better relationship with your cat, find
the sections that pertain most closely to your situation.

   If you do not have a cat but have a strong interest in tigers and
large cats, you can read just the tiger stories in each chapter.

   We, as animal behaviorists, work to create a more stimulating
and healthy environment for animals and to develop our own understanding and awareness of our pets' behavior and how we can live
with them in harmony. We are very pleased to share our knowledge
and techniques.

close cousins

*Prowling his own quiet backyard or asleep by the
fire, he is still only a whisker away from the wilds.*
Jean Burden

A WHISKER AWAY FROM THE WILDS

His body melts into the ground, his eyes blazing with a primitive excitement. His tail twitches with a hypnotic effect, the body tense with each muscle rippling, storing more energy by the second. He looks like a statue coming to life. Slowly one paw creeps forward, then the next, his body compressing into a spring ready to explode. Suddenly the explosion happens . . . all thirteen pounds of Fridge the tomcat pounces into the air and a young sparrow narrowly escapes.

The effect of a cat whether it's a five-hundred-pound Bengal tiger or a ten-pound domestic cat stalking its prey is the same. The motion is intense, powerful, and hypnotic. The similarities between them by no means stop there: The behavior of a tiger and a house cat is extremely similar. Take away the size difference and markings, and a cat is a cat! Our experience living and working with tigers has given us a new appreciation and understanding of our domestic cats. In the ten years Bill has been working with Bengal tigers and we have been consulting people on their relationships with their house cats, we have found the parallels between cat and tiger remarkable.

Many people are surprised to discover that tigers are not just fearsome and aggressive hunters. These strong cats can be exceptionally playful when interacting with each other—or with a human they trust. They sleep in precarious positions and chase their own tails. And cat owners everywhere soon learn that their house cat has a side that shows that there is a touch of his wild cousins in him that has never been left behind.

PHYSICAL SIMILARITIES

OF ALL THE DOMESTICATED ANIMALS IN THE WORLD, THE CAT HAS BEEN DOMESTICATED THE SHORTEST DURATION OF TIME.

●

People always ask questions about the physical likenesses between a tiger and a cat. There are many more similarities than differences, like the purpose of color and markings. For both cats, they are camouflage. It's difficult to believe that the tiger's orange and black stripes are a perfect camouflage for him in his natural environment. (Every newborn cub, weighing in at a beefy two pounds, comes with all his markings and stripes. The striped pattern is like a fingerprint and will not change throughout the tiger's life.)

In many settings, a tiger will stand out like a piece of art, but in the wild, she can slip away and disappear in the blink of an eye. With her light underbelly, the tiger blends naturally into the shadows. The stripe pattern breaks up the contour of her body. Bill is frequently surprised when a tiger seems to disappear from where he saw her just moments before. The tiger didn't move, but she's gone. Bill will walk through the grass looking for her . . . then *Boom!* he's on the ground with paw prints up and down his back—being playfully abused by a sneaky tiger.

All the different shades and colors of domestic cats are also

camouflage. But with selective breeding and natural mutations, the house cat now comes in colors not likely to be found in the wild. Still, the tabby coloration found in a variety of cats is a good example of stripe patterns breaking up the contour of the body, which helps them escape detection. Even marbled, patched, and tortoiseshell markings found on cats give them the ability to blend into their outdoor surroundings and help them slink around unnoticed. Both tiger and cat camouflage vary according to the original environment from which the cat came. Stripes and spots blend in with the surroundings in the forests and jungles, tawny and sandy colors in semidesert regions. Even the white tiger has camouflage: The stripes break up the form of his body making him difficult for prey to spot before it's too late!

The purpose of tigers' and cats' beautiful coats is also the same: to protect them from the elements. But the feel of their coats is quite different. The house cat's fur is soothingly soft to the touch, whereas tiger fur has a much coarser feel, similar to that of a large dog. Tigers' fur is extremely thick because of their size and because the

TIGERS ARE THE BIGGEST OF ALL CATS. THE SIBERIAN SPECIES, WHICH CAN WEIGH OVER 650 POUNDS, IS THE LARGEST.

•

area from which all tigers originate—Northern Asia, where the massive Siberian tiger still exists in the wild—is extremely cold. Tigers migrated all through Asia even to the southernmost region of Bali, where the tiger is now sadly extinct. (The last tiger seen on this exotic Indonesian island was in the early 1940s.) Tigers became smaller in size as they migrated south and their coats became thinner to adapt to the warmer climates.

Tigers and cats both groom themselves, though tigers are not as meticulous as cats. Cats spend a great deal of time cleaning themselves and even avoid dirt. They tend to clean up immediately after a confrontation with any messy substance they encounter. Tigers don't mind being a little dirty now and then and often procrastinate before grooming. Tara, the most talkative of our tigers, runs along and stumbles into the mud, but does she stop to clean herself? Nope, she takes it in stride and keeps going. If powerful Mohan is spraying urine and Baghdad walks in the way, does she mind? Not a bit; she just continues on with her business. Although tigers do spend a good amount of time cleaning and grooming themselves, they also indulge in playing in the dirt and don't bother to tidy up for hours.

Whereas most house cats needing a bath never go voluntarily, a tiger's bath is a swim or run through a river or pond. When finished, he shakes off the excess water, then settles down to finish with a "once-over" tongue bath. The last thing left to do is a warm sunning to dry off. Ahhhh, the life of a tiger.

The tiger's large tongue, which is so good for grooming, is similar in texture to the tongue of her domestic relative. Both are rough because of the hundreds of papillae pointed in a backward direction over the surface. In fact, a tiger's tongue is so rough she can lick

WITHOUT THEIR FURRY COATS, LIONS AND TIGERS WOULD NO LONGER BE ANATOMICALLY DISTINGUISHABLE. THEY CAN MATE AND PRODUCE CUBS THAT IN TURN CAN REPRODUCE.

•

THE CAT HAS 30 TEETH, FEWER
THAN ANY OTHER CARNIVORE
(MEAT EATER). HE ALSO HAS THE
HARDEST BITE BECAUSE OF HIS
SHORTER MOUTH.

●

the paint off the side of a house. Using her tongue, a tiger can completely clean the meat off a bone from prey she has hunted. People who know this always get wide-eyed as they're watching a tiger lick Bill's hand. But tigers have good tongue control and apply only as much pressure as needed. We've seen a mother whose tongue was bigger than the cub gently cleaning her newborn. And we've seen Mohan so obsessed with the scent on a rough piece of wood that he licks it baby-bottom smooth, as though his tongue were sandpaper. On a hot day, Mohan likes to come over and lick the salt off of Bill's exposed skin—Mohan's large tongue easily covers Bill's forearm with one lick. Occasionally when Mohan gets carried away with licking, he

licks with the same vigor but, getting tired of standing, puts more of his weight into the lick. When Bill's skin starts to be pulled too hard and becomes red, this is the time to say no—or switch arms.

All cats and tigers do what is called a Flehmen response, which looks like a giant smile. The nose wrinkles, the tongue hangs out, and they grimace. (This is a great time to see the tiger's impressive canine teeth.) The reason for this response is to allow them to record a scent by breathing it into a sensitive area in the back of the mouth called a Jacobson Organ.

The whiskers that stick out from tigers' and cats' cheeks, called vibrissae, are used as a sensory device. When an object makes even ever-so-slight contact with the tip of a whisker, a signal is sent to the sensitive nerve endings at the whisker's root. A message of the contact is then relayed to the brain. Whiskers help guide tigers and cats as they walk in total darkness, giving them the ability to detect even the slightest change in air currents around objects.

Tiger whiskers feel just like thin quills. They are black near the fur and change to bright white toward the ends. Kali, a spunky female tiger, has whiskers that really enhance her striking looks. Her long, elegant whiskers by far surpass those of other of our big cats, and some of them reach out over seven inches. Cats can also have whiskers as long as Kali's and just as striking, but softer in texture.

Tigers' and cats' sight is their most acute sense; they can see extremely well at night and in the dimmest light. The reason cats' and tigers' eyes turn luminous green at night when a light shines on them is because of the reflective *tapeturm lucidum* (a layer of flattened nerve fibers) located behind the eye. This is made up of fifteen special glittering cells that use the bright color to magnify the light.

A tiger's pupils contract into a round shape, the same as a human's do. A cat's pupils contract into a vertical slit. Cats' eyes are more sensitive to light than tigers' and they are able to block out more light with the vertical contraction. Cats tend to be even more active at night than tigers so they need this extra sensitivity. Both tigers' and cats' eyes dilate into a round shape and both have sight that is six times stronger than that of human eyes in dim light. All tigers' eyes are a golden color, except for the white tigers', which are chilling blue. House cats' eyes vary from sapphire blue through many shades of green to gold.

Another physical similarity is that all cats have certain glands that let them mark their territory by rubbing, clawing, and spraying their individual scents. This marking by scent is a key social activity. Both males and females spray urine as a kind of calling card. (The odor of tiger spray is like stale popcorn.) Most people are surprised to see female tigers spraying, and even more astonished when told their own female cat can spray. The scent can identify the individual, the sex, whether a female is in heat, and possibly even the tiger's age. Tigers will mark and remark their territory to signify and update their presence. They have excellent aim and can spray up to fifteen feet. Watch out if a tiger ever raises his tail to you—even if there is distance and a fence between you.

One of the differences between a tiger and cat is the kind of sound it is able to make. Members of the large cat family, including the lion and tiger and jaguar and leopard, cannot meow or purr. Big cats are most famous for their mighty roars, but they can also make several other sounds like growling, moaning, groaning, and a sound called a chuff.

HUMANS HAVE A VISUAL FIELD OF 210 DEGREES, OF WHICH 120 IS BINOCULAR OVERLAP (BOTH EYES HAVING THE SAME FIELD OF VISION). DOMESTIC CATS HAVE A TOTAL VISUAL FIELD OF 285 DEGREES, OF WHICH 130 IS BINOCULAR OVERLAP.

●

FELINES IN GENERAL WALK ON
THEIR TOES, WHICH ALLOWS
THEM TO RUN FASTER THAN IF
THEY WALKED ON THE FLAT OF
THEIR PAWS.

●

If you put your lips in the position to make the *f* sound, and you blow very hard in quick intervals, the resulting *f-f-f-f* is what a chuff sounds like. It is a very important sound for tigers, as it is their way of greeting and showing their moods. A trainer must be able to chuff back, returning the greeting. If the trainer chuffs and does *not* get a response, then he knows something is up—and to be careful. Most of the time a tiger will respond with a hearty chuff which can be seen even when too far away to be heard, as the tiger's head tends to bob up and down. When chuffing up close, tigers usually like to rub their heads from side to side on any convenient body part of the trainer, usually his face. Like the sound of a contented cat purring, the chuff of the tiger warms the heart.

Cats, who are able to meow, cannot roar or chuff. The reason for this difference is that the roaring big cats—tigers, lions, jaguars, and leopards—have elastic sections on either side of their hyoid bone (the bone that supports their tongue). These elastic sections enable the vocal apparatus to move freely so that these wild creatures can let out more vigorous sounds. In smaller felines, the hyoid bone is solid and not able to stretch, preventing them from being able to roar like their ferocious-sounding relatives.

Another difference between the two is their climbing abilities. Most domestic cats are agile climbers and never have problems until it comes to getting down. Tigers have difficulties with both up and down. When tigers are young they experiment with climbing just as all cats do. Cubs are quite uncoordinated and will try jumping up about five feet and latching onto a tree. Finding themselves in an awkward position, they immediately become nervous (and a vulnerable target for an ambush from their littermates) and will get excited,

DOMESTIC CATS HAVE A HARD
TIME GOING *DOWN* BECAUSE ALL
OF THEIR CLAWS POINT IN ONE
DIRECTION. ALSO, THEIR TOES
AND MUSCLES AREN'T AS FLEXIBLE
AS THOSE OF OTHER TREE
CLIMBERS.

•

jump off, and run away. Tigers do not know they aren't expert climbers like their little cousins and need to learn this for themselves.

Most people don't realize the passion a tiger has for the water. This is the other distinct difference between tigers and cats. Tigers will spend countless hours submerged up to their heads in water—to combat the heat or just in a frenzy of splashing play.

TIGERS ARE ONE OF THE TWO
LARGE CATS THAT ACTUALLY SEEK
OUT WATER.
THE OTHER EXOTIC AQUATIC CAT
IS THE JAGUAR.

•

TIGER ISLAND

Tiger Island isn't an island at all, but an expansive grassy area with several trees, including an enormous willow tree which provides shade, and a pond. This area, originally built in 1975, is a very special part of Marine World Africa USA in Vallejo, California. It has several large logs and boulders strategically placed so that a tiger can jump from one to the other. Tiger Island has a deep surrounding moat to keep a safe distance between the public and the tigers. Ten tigers roam the lush environment, roll in the shade under the cascading limbs of the willow, or sleep and hide in tall emerald-green grass. The large logs there are not only good for jumping, but they give the tigers a higher vantage point, which is also a favorite sunning spot. The placid water in the pond is extremely inviting to the tigers on a hot summer day.

One of the most popular female tigers on Tiger Island is Baghdad, who has not only stolen a big piece of my heart but is also a big favorite with the general public. She has received a lot of press and is the star of several posters and prints. She is in the *Animal Guinness Book of World Records* for having had the largest recorded litter of cubs. In 1979 she gave birth to eight cubs of which seven survived. For Baghdad, this made a grand total of forty-three cubs, all born on Tiger Island.

Tiger Island has played an important role in preserving the tiger, which is on the endangered species list, by educating the public on the crucial need to protect the tiger and its habitat, and because of the research on tiger behavior that takes place there.

There is no other place like Tiger Island in the world. Have you ever seen an exhibit where a tiger is pacing endlessly or staring blankly, looking despondent and bored? On Tiger Island you actually see bright, excited eyes, and an unleashed exuberance—the same spark of life you see in a happy house cat.

I have been training the tigers on Tiger Island and with the help of the other trainers have created a progressive training program that keeps the tigers stimulated while under control. It is a program that relieves any boredom or stress the tigers might otherwise encounter and enhances their well-being. Tiger Island is not only stimulating for the tigers, it's also a very exciting place for people to watch tigers behave naturally. Instead of just walking by an area where a tiger is sleeping, people stop to watch them and take the time to ask questions of the trainers about what they can do to help protect the tiger.

THE TIGER ON YOUR COUCH

As you can see, there are many similarities between tigers and cats, both behavioral and physical. But the best teacher of cat behavior has been Baghdad. At seventeen years old, she is a hefty four hundred pounds, which is a hundred pounds heavier than the average female. She is a yellow-orange color with a short, silky coat and just a little ruff around the sides of her face. She has a soft look in her warm golden eyes. Baghdad also has a unique greeting habit all her own. She will come over in her usual way with a friendly chuff, then suddenly jump up and wrap her paws around your shoulders, very similar to a hug. At the same time she rubs her head from side to side on your face. This kind of greeting is so powerful that many times I have found myself on the ground when I wasn't ready for a Bag hug.

It took three years and many sharp looks and snarls before I truly began to understand her and feel comfortable around her. In those years, after spending consistent and quality time with her, I finally earned her respect and trust. Once you've won Baghdad's trust, she gives an enormous amount of warmth and affection. It took a lot of patience, but the respect and trust that developed is mutual and lasting.

THE AVERAGE LITTER FOR A TIGER IS 3 TO 4 CUBS, AND FOR A DOMESTIC CAT 6 KITTENS.

•

Mole was a house cat that we adopted from the pound. She stole our affection and taught us a thing or two about cat behavior. Mole was a cat that blended into her surroundings—with a quiet presence that made her go unnoticed when company came over. Although she was quiet, her eyes told all. We learned that eye contact

THE WHITE SPOTS BEHIND
YOUNG TIGER CUBS' EARS
SIMULATE EYES AND MAKE THEM
APPEAR LARGER TO A PREDATOR.
●

was the only directive she needed; that would bring a soft rub, a quiet meow, and a look that would melt the coldest of hearts. She would then settle in right next to us, back to her quiet ways without another meow or demand for attention, just a gentle purr. Spending quality time with "the tiger on your couch," and conditioning behavior consistently, will be equally rewarding and allow the two of you to have a happy life together.

choosing your feline companion

CHAPTER 2

*I love cats because I enjoy my home; and little by
little, they become its visible soul.*
Jean Cocteau

DOES YOUR LIFE-STYLE FIT A CAT
AND DOES A CAT FIT YOUR LIFE-STYLE?

When you are deciding to take a cat into your life, the most important
question to ask yourself is whether you have quality time to commit
to a cat. Long hours aren't necessarily the key—quality time spent
together is. It's easy to think that a cat is a low-maintenance pet and
assume she can fit into any schedule. Many people think that cats are
independent and like to be alone. The truth is that cats may not be
demanding of our time, but they are very much in need of it. Sure,
you don't have to rush home and walk them the way you do with
dogs; but cats have strong emotional ties to their owners. If your job
has long demanding hours or you tend to travel a lot, you may want
to wait until your life-style is more settled.

   Each year, thousands of cats are abandoned or given up to
animal shelters because of changes in people's life-styles: job trans-

fers, moves, relationship changes, increased hours at work, and a variety of other situations. People don't take the time to find proper homes for these cats and they are left to wander the streets or be picked up by the local animal control. The decision to take a cat home is a serious matter. On the average a cat can live fifteen years when properly cared for. Your commitment will need to be firm and steadfast.

If there are going to be occasions when you must travel or be away from home, the question to ask yourself is will you be able to find a good kennel or someone who will care properly for your cat. It needs to be someone who will do more than just put out Kitty Litter, food, and water; someone who will truly take notice of your cat's physical and mental well-being. The tigers on Tiger Island have a trainer with them 365 days a year. If one of us goes on vacation, we know there are three other trainers who will give the big cats the

THE FIRST TAMED CATS WERE USED FOR PEST CONTROL IN ANCIENT EGYPT ABOUT 3000 B.C. AND CAME TO BE LOVED AS HOUSEHOLD COMPANIONS AND WORSHIPPED AS GODS.

•

quality of care that they need. This system isn't realistic for all of us to use with our house cats, but before "jumping in with all paws," make sure you have a reliable kennel or friend who will have your cat's well-being in mind. A friend who can stay in your house is best. If you need to leave your cat behind, it's always better for the cat to stay in familiar surroundings. Fewer changes at one time mean less stress for your feline companion.

Another point to consider before bringing home a cat is whether or not your pocketbook can handle it. There are expenses in keeping these household companions healthy. Besides daily food, toys, and Kitty Litter costs, there will be the initial costs of food and water bowls, a litter box, scratching post, and carrying box. Remember too that visits to the veterinarian will be necessary. The costs will include vaccinations, checkups, spaying or neutering, treatment if injury occurs, or medication when needed. It's easy to help keep your vet bills minimal by being constantly aware of your pet. You can be your cat's best vet (see Chapter 8).

Do your living quarters have enough space to be shared with your new companion? You obviously don't need the same space as you do for a tiger, or a golden retriever for that matter, but your cat will need enough room in which to romp around and exercise. If you don't live alone, how does the rest of the household feel about having a new member of the family? Is anyone who lives with you or frequents your house allergic to cats? These are always important questions to answer before you choose your cat.

If you ever want to brighten someone's life with a kitten, make sure he or she *really* wants the responsibility first.

We will categorize some of the less glamorous aspects of being a cat owner, just to paint a complete picture before you make this big decision.

- **Litter Box**   Inside cats, outside cats, all cats need a litter box. It's not so unpleasant, but it is a chore that you or a household member will have to do *daily*. The tigers' litter box is as big as their grassy environment, so Bill's daily chore at work is quite a bit bigger than at home.

- **Hair Shedding**   A cat's shedding her hair in your house is part of life with a cat. Long-haired, short-haired, all cats shed. You can minimize shedding by brushing your cat daily, which can be relaxing and enjoyable for both of you. Judy's theory on the subject is "if you have a light carpet, then select a light-colored cat, and if you have a dark carpet, choose a dark-colored cat."

- **Kittens Grow into Cats**   Kittens are adorable little balls of fur who grow up into distinguished cats. I've heard comments from people who say that "it would be great if they would stay in the kitten stage." Personally we couldn't disagree more. We completely enjoyed all the tigers as cubs, and our cats as kittens, but we have grown together so much through their adulthood.

- **Neighborhood**   Check your neighborhood out to see if there are many dogs that are allowed to roam freely. Also check out how many tomcats are around who haven't been neutered, especially if you are thinking of getting a male. Unneutered tomcats are typically territorial and will pick fights with other cats.

- **Landlord**   If you rent, be sure to look at your lease. If there isn't anything about pets in writing, let your landlord know your plans. You may have to put down a pet deposit, but never try to sneak a cat into a residence. Landlords are usually more cooperative when you're up front with them.

- **"Oops, My Favorite Vase"** Any kitten or cat can accidentally knock things over. Material items can be repaired or replaced, but the affection of a cat is irreplaceable!
- **Not Always Silent** Cats can be quite loud and have a distinctive meow. As cats tend to be more active at night, the pitter-patter of scurrying feet or meowing outside the window are possibilities.

Some breeds of cats are louder than others. Siamese, for example, are known to be very vocal. Other cats are loud by character. A good example of a loud tiger is Tara, who can make a Siamese seem quiet! She makes a variety of sounds that Bill has never heard from any other tiger. Tara constantly moans, groans, chuffs, and grumbles. She may be peacefully lying down and can still be heard from twenty yards away. During so-called quiet times with her, when she is being showered with rubs, pats, and scratches, there are nonstop vibrating noises coming from her, some of which sound just like a violin badly out of tune.

There are so many factors to think about before bringing home a pet. When deciding to make a commitment to a cat or kitten, remember to take a hard look at your life-style to determine if a cat can fit into it. Also take a good look at the needs of a cat and decide if your life-style can meet them. The fulfillment the presence of a feline brings into a home and your life, in our eyes, outweighs any inconveniences, ten to one!

ARE TWO CATS BETTER THAN ONE?

It's natural for cats to stick together during the formative years. There will usually be two to four littermates who turn into squirming little playmates. They learn to play, investigate, and develop their hunting skills together. In the wild, a tiger cub will stay with his mother and siblings for two years. This time is of a much shorter duration with domestic cats but is very important in their social development. It's natural for a kitten to have companionship; this gives the kitten a constant source of stimulation.

If you or one of your family members is not home continually to provide social growth and relief from boredom for your kitten, it is best to have two cats. They keep each other company, play, and have each other's warmth when they sleep. Even if you are home most of the time, two cats still are better. Kittens are a bundle of energy. In rambunctious play with their littermates, they are able to unload their energy, which then helps them relax when interacting with you. Kittens and tiger cubs both play extremely rough games with their littermates, which means a lot of biting. However this type of play is not permissible with you, because as with the tiger, it will lead to many problems when they're older. Letting kittens play in a fury of bumps, bangs, jumps, nicks, bites, and scratches is great for their development. When playing with you, though, there are definite kitten rules that need to be set to avoid problems later on (see Chapter 3).

There is a misconception that if you have two kittens they will be more interested in each other than in you. At first, this will be true because they act as support for each other in new and scary situations. But if you spend time with your kittens and are consistent in your attention to them, they will soon come to you for affection and protection, as if you were their mom (see Chapter 3).

There are other advantages to bringing home two kittens or cats:

- They will have companionship in your absence.
- They will develop more confidence with the support of a buddy.
- They can take their energy out on each other, saving wear and tear on you and your household items.
- There is minimal difference in responsibility in caring for one kitten or two—but there is twice the joy.

Are littermates a good choice? Unless you plan on breeding your cats, the answer is yes. Although you will have to spay or neuter them, which we recommend highly in almost every case (see Chapter 10), littermates tend to be a little closer and are more forgiving of each other: Starting their relationship from the first days of life, they have often built a strong bond.

The tigers are exactly the same in this. Kali and Rakon, two eight-year-old tigers, are from the same litter and are still great playmates. Kali gets along the best with Rakon of all the tigers; she will happily sleep next to him and occasionally groom him. Serang is the same age but from a different litter. Kali will completely take advantage of his laid-back ways and Serang is not so watchful of her sneaky nature. Kali has been known to torment Serang several hours at a time, sometimes on a daily basis.

Often it isn't possible to house two cats, because of a small space or limited funds, for example. It may also be easier for an elderly or differently abled person to care for only one cat. If you can't or don't get two cats, the best solution is to spend as much time as possible at home, giving your one new friend tremendous amounts of affection!

CHOOSING A KITTEN OR CAT

Adult cats: There are many advantages and one disadvantage in selecting an adult cat. If you are able to spend time with the cat beforehand, as when you sit for a friend's cat, then you have the advantage of seeing what you're getting. You can usually find out about the cat's history and any past health problems. You can learn about the cat's temperament, which will help you to determine if this cat is a good match for you. You can see what specific traits the cat has, such as whether it is nervous, noisy, aggressive, a people lover, a plant eater, a loner or gets along with other cats, and so on. This will help you make a smooth transition with and understand your new pet so you may quickly develop a good relationship.

An adult cat will be less demanding and need less supervision in the initial stages than a kitten. Find out if your future companion has been taught litter box use, the word "no," and other conditioning patterns. These have probably already been established if this cat has had an owner-to-pet relationship.

Think about taking in an outside stray. It's a pretty sure bet that the cat is already streetwise, and if she has a good temperament, she can make a great pet. Orphaned strays can often be real charmers

and capture your heart. Another point to think about when determining if you should select an adult or not is your house. Look around to see if you have expensive items that an out-of-control kitten could damage. Mature cats tend to be less reckless around the house than kittens.

The only disadvantage in getting an adult cat is you are not able to shape her personality while she is growing up. If the cat has grown up in a home without affection, she may tend to be less affectionate with you; if she has grown up getting by with many bad habits, they will be harder to change.

Kittens: If you want to influence the development of your new pet's personality, then it's best to start with a kitten. You can condition a kitten to follow the guidelines and rules of the house; you can nearly condition a kitten to your own personality if you put in the time. If you want a cat to behave like a dog and greet you at the door, come running when you call, and take walks on a leash with you, this is possible if you take the time. If you want your cat to be more

independent, affectionate, gentle, whatever, it's all possible when you start conditioning consistently during his kittenhood.

Another advantage to getting a kitten is being able to watch him grow up through all the different phases of his life. We've learned so much from the tigers and our cats by watching them mature—every phase is entertaining and educational.

Kittens, though, do need more supervision and time than mature cats. They will require several meals a day, so someone must be there for them. From eight weeks to three months old, they must be fed four times a day. Then from three to six months old, they must be fed four to six times a day. Kittens are also more destructive. Those little balls of fur can get in more trouble than you can imagine! The American essayist Agnes Repplier captures a kitten's spirit by writing, "A kitten is the most irresistible comedian in the world. Its wide-open eyes gleam with wonder and mirth. It darts madly at nothing at all, and then, as though suddenly checked in the pursuit, prances sideways on its hind legs with ridiculous agility and zeal."

If you already have a cat and are considering a second one, a kitten will be less threatening to your present pet.

HEALTHY SIGNS

Once you've determined that your life-style can accommodate a cat, the next step is to decide what kind of cat and where to get it. The most important thing you should consider when looking for a kitten or cat is health. You will want to give your potential pet a solid look-over before deciding. Here's what to look for:

- **Eyes** should be alert, quick to react, and have no discharge.
- **Ears** should be clean and show no signs of infection.
- **Fur,** the overall coat, should look healthy and feel soft to the touch. Check for sores, dry patches, hair loss, infestation, and scabs and dirt left from fleas.
- **Mouth**—look inside and check teeth and gums. The gums should appear a nice healthy pink.
- **Nose** should be clean and have no discharge.
- **Breath** should not be labored and should sound clear.

Remember, no matter how healthy your new cat or kitten appears, she will need to be seen by a veterinarian and receive all her vaccinations and a clean bill of health.

If you find a cat or kitten (or she finds you) who is *not* healthy, it is a very rewarding feeling to nurse her back to health. We found an abandoned litter in a field one time. Once the kittens were put on a nutritious diet, treated for fleas, and received their shots, they were healthy enough for us to find suitable homes for them. We still get reports back on what wonderful pets these little orphans made. Nursing an unhealthy new pet takes much time and the expenses can add up, but it can be very fulfilling and bring you even closer to your cat.

Two very lively examples of the rewards of committing time and effort to helping disabled animals are Kehei and Esi, two wonderful and very special tigers. Esi and Kehei are physically handicapped and unable to function without human assistance. Esi is partially paralyzed in her hindquarters, having only 50 percent mobility in her back legs. Kehei was born with a genetic eye problem. Barely able to see as a cub, he gradually lost the rest of his sight. He is now completely blind and must rely on his trainers as his eyes.

Because of Esi's and Kehei's handicaps, they have become extremely dependent on me and the other trainers, which has created a special bond and trust between us. Although Kehei is blind, he still leads an exceptionally full life and is comfortable with his surroundings and trainers. I take him out to the grassy field on a leash and then let him roam free and behave as the rest of the tigers do. Kehei is very familiar with his surroundings, and like the other the big cats, he walks around and investigates every bush, tree, and scent. Once finished with the investigation, Kehei sprays to mark his own territory. He knows the boundaries of Tiger Island and rarely bumps into anything.

Tigers rely extensively on their sight, so it's amazing to watch Kehei as he maneuvers so well using the rest of his senses. He constantly stops, lifting his immense head to smell in all directions to sense what's going on. You can imagine how busy he stays on a windy day. He also moves his head from side to side focusing his acute hearing, taking note of what each noise is and from what direction it's coming. Just like Rakon, Kehei knows the sounds of food in the making and always comes running over. Through the years he has adapted so well and has seemed so normal, it is extremely easy to forget he's blind.

We often take Kehei for walks on a leash outside of Tiger Island to stimulate him with new surroundings and scents. When I gently tap the ground with a wooden stick to show him changes in terrain, Kehei knows this means to slow down. He then reaches out with his paw to feel if he must step up or down. As we have already described, tigers can be unduly excitable, especially when frightened, reacting first and thinking after. Because of Kehei's total reliance on us, we have developed his confidence to listen to us and trust that we will help him in all situations.

Kehei loves the water and spends hours splashing away in it. As with the other big cats, we frequently swim with him. On hot days we often bring out toys for the other tigers to chase and play with in the water, but these toys can confuse Kehei and make him misjudge what he is landing on. On several occasions, Kehei's confused landings were right on top of trainers. After one too many of these rough landings and enough abuse to our bodies, we knew we had to invent a fun and safe toy for Kehei. Finally we tried using a high-pressure hose to spray into the pool. To our great joy, this device was a tremendous hit. He gets excited as soon as he hears the noise of the hose and quickly feels for the bubbles made from the water pressure. He starts chasing the water and biting at the spray, all the time making a tremendous amount of splashing.

Kehei takes short breathers from the fun but is keen to let you know when it's time for another hosing. It's very rewarding to watch Kehei enjoy these playtimes, since it's a good reminder of the confidence he has built in us—he is comfortable enough that he can relax in many different situations. I feel fortunate to be trusted by Kehei and to have shared so many special moments with him.

Esi had a long and difficult road to recovery from the accident that paralyzed her. Several years ago on a cool winter morning the grass was wet from a heavy rain. Esi was running along, happy to be wet and pleased with the cooler temperatures. She spied a trainer and started running toward him to tackle him. This was an amusing activity for her, but not as much fun for us. As she jumped up her hind legs suddenly slipped out from under her, causing her spine to twist. It was a freak accident and instantly Esi lost all feeling and mobility in her back legs. Though she was shocked and confused, Esi showed no signs of aggression. To react this way when injured is very uncharacteristic for a wild animal; it showed that Esi completely trusted the trainers to help her. This trust allowed us to take her to comfortable quarters where we could work closely on her. At that time the veterinarian, Lauri Gage, could not predict Esi's future. She could only recommend that we try to help her as much as possible and wait to see the outcome.

For four weeks we spent long hours massaging, stimulating, and moving her legs for her. She finally regained some feeling and started moving her legs on her own. We would slowly help her to her feet. She was a bit wobbly at

42

first, then she began to walk. Esi made daily improvements as we helped her up to walk, trying to rebuild her muscles that had atrophied.

Our hard work and her persistence paid off. Esi now moves about on her own and on hot days she can be found energetically swimming about in the pool. Though she swaggers back and forth when she walks, Esi is still capable of a quick dash. She will never be totally functional and sometimes needs our assistance when she's tired, but Esi is in no pain and she has the heartiest of chuffs and the warmest of greetings. Most of all, she shows that spirit in her eyes, which is the true mark of a happy tiger.

Both of these animals owe their happy existence to the team effort of several devoted people and to their own wonderful characters. Because of the unique training we did, the relationship we developed with these two tigers, and the determination of the veterinarian staff, we were able to find answers and solutions to help them.

There is little that can replicate the joy of helping and living with a cat you have nursed back to health or a disabled cat whose pain you have relieved. The satisfaction I feel from Esi's and Kehei's trust and affection is unmatched by any other feeling I've ever had.

THE FIRST CAT SHOWS STARTED
IN EUROPE.
THE FIRST U.S. CAT SHOW WAS
HELD IN NEW YORK IN 1895.
●

GETTING THE PICK OF THE LITTER

Eric Gurney, poet and cat lover, observes, "The really great thing about cats is their endless variety. One can get a cat to fit almost any kind of decor, color, scheme, income, personality, mood. But under the fur, whatever color it may be, there still lies, essentially unchanged, one of the world's free souls." When selecting your cat or kitten, it's tempting to pick the cutest one, which is fine if your first impressions are usually dependable. A good rule of thumb to follow that works well is go for personality over looks. Choose a cat who seems outgoing and has a good temperament. Of course, cats being cats, at any particular moment a cat may be hiding his true character. He may have just come into a shelter and be frightened. Or maybe it's not his day or he just wants to take a nap. Take some time and draw out the cat's personality. Here are some pointers:

- If there's more than one cat, observe from a distance to see how they interact with one another.
- Get close and watch for a reaction . . . then reach out and let them come to your hand. This will show you how brave they are.
- Pick them up in a slow, gentle manner and see how they react to being touched.
- Bring a toy, something like string or ribbon, something small and nonthreatening, to play with. You're looking for a playful reaction, not to scare them.
- Give all the kittens or cats a chance, not just the first one who greets you.
- Take the one you think you might want away from the group and spend a few minutes together. Being away from the others often brings out the cat's personality.

WHERE TO FIND YOUR FELINE

Shelters: Your local animal shelter, or animal control department is a wonderful place to pick a cat. We have gotten two cats from the pound, Mole and Penguin, and they have made wonderful pets. Animal shelters have many cats to choose from, with kittens being the most abundant in the spring and summer. In most cases, the kittens or cats will have a friendly disposition. (Sadly, many more kittens and cats come to these organizations daily than can ever be adopted, so only the healthiest and best-tempered cats meet the public.) If they're over six months old they will have already been spayed or neutered, and most have been litter box trained. And your potential companion will have been given a complete medical examination. In addition, we've always found the shelter people extremely helpful. They will assist you in making a good choice and provide any other advice you may need, such as where to take your cat to update vaccinations. You can look in the Yellow Pages for the nearest animal shelter.

Cat Breeders: Picking a pedigree cat is not easy. There are over a hundred different breeds and color varieties. The only reason to pick a pedigree pet with so many cats in abundance is if you are attracted to a certain look or personality trait.

When selecting a breeder, do some homework and make sure you find a reputable one. Go to a cat show and talk with the owners and breeders. They will be able to help direct you to breeders in your area. Be aware that choosing your new cat through a breeder can be expensive. You also need to know that there can be many health problems resulting from excessive inbreeding.

These health problems often occur with white tigers such as Sampson. Sampson is a beautiful tiger with ice-blue eyes and black stripes set off in a field of snowy white. The normal coloration for a Bengal tiger is red-yellow to red-orange; Sampson is an off-color variation, a kind of variation that occurs in many animals. Most people think he is an albino. He is not. You can tell the difference by looking at the eyes; a true albino always has pink eyes. There is no mistaking when looking into Sam's piercing blue eyes! You can also tell by his black stripes; an albino is an animal with an absence of pigmentation.

Sampson is a robust and healthy young white tiger. White tigers are the result of a recessive gene which must be watched carefully through selective breeding. Sam's heritage goes back to another tiger by the name of Mohan, the last-known wild white tiger who was found in 1951 as a nine-month-old cub. Selective breeding is crucial for the health of a white tiger. If too much inbreeding is allowed, the results can lead to several health hazards including eye, hip, lung, and abdominal problems.

NEVER TAKE A KITTEN FROM HER MOTHER BEFORE 8 WEEKS. A KITTEN NEEDS THIS TIME FOR PROPER DEVELOPMENT.
●

An advantage of going through a breeder is that you are able to take a good look at the whole litter to choose the one kitten you favor. If you are picking a kitten, closely observe the litter and see how the mother is interacting with them. Also take note of the environment in which the cats are living. Ask questions about the mother's personality. The care given by breeders is usually good and it is standard that they be examined and given shots by a veterinarian.

Stray Cats: As in many aspects of life, what happens to us doesn't always come down to choice. Many times cats come into our lives by chance or circumstance—they may be lost, hurt, or just in need of a home. Writer Eve Merriam expresses it best: "It's just an old alley cat that has followed us all the way home. It hasn't a star on its forehead, or a silky satiny coat. No proud tiger stripes, no dainty tread, no elegant velvet throat. It's a splotchy, blotchy city cat, not a pretty cat, a rough little bag of old bones. 'Beauty,' we shall call you. 'Beauty, come in.'"

We adopted a stray cat while living in East Asia. After a lot of nurturing time, we developed a mature and respectful relationship with her. If you are a kind soul who takes in a stray, remember that you won't be able to tell about her past, whether she was the spunky one of the litter or the quiet one. A stray may have picked up some bad habits that cause some trouble down the road. Always take a stray cat to the vet immediately for a complete checkup and shots. This is a must particularly if you have other household cats because there are many transmittable diseases that can cause health problems. You should also isolate the new cats from your other cats for ten days to help make sure your new cat does not bring any diseases to the cats you already have. With a stray cat, your relationship needs to start out with kindness and can soon develop into an inseparable friendship. Both of you will need to learn about each other's habits, and you need to teach her the rules of the house, which will take time and patience on your part.

Because tigers originate in Asia, it is geographically impossible to pick up a stray, though several Nepali villages have picked up stray orphan cubs to nurse and released them back to the wild. Bill has always worked with the cats born on Tiger Island. It wasn't as if he could work with just "the pick of the litter"; there was no choice. He now loves them all. What we've learned is no matter what the tiger's or cat's disposition is, if given a chance and some caring time, any cat can make a wonderful companion.

TIGERS AS PETS?

As the French writer Fernand Mery said so well, "God made the cat in order that man might have the pleasure of caressing the tiger." People have owned exotic cats in the past, but finally laws have been enacted making it against the law to own any endangered species. (And it's very sad that the tiger is an endangered species.) Only accredited facilities with the proper environment—which includes ongoing supervision, a full-time veterinary staff, and an invigorating food bill—are allowed to house tigers.

When people owned tigers as pets, the tigers almost always suffered from poor physical and living conditions. People would fall in love with a cute tiger cub with large paws and pretty stripes, but a tiger is not a house cat and he grows up quickly. He soon starts to bite and scratch if not properly conditioned. In some cases, private owners declawed the tigers in an attempt to control them. This did not change the tiger's temperament and caused great discomfort. The majority of the time, the tiger ended up in a small cage because the owner could not control him any longer and realized it was not safe to try.

We've often been asked if a tiger or any other big cat makes a good pet. The answer is always *no*. Bill has been living around tigers almost half his life and loves them dearly, but they are not meant to be kept as pets. Working with them is a full-time commitment. A tiger as a pet is not practical, nor is it fair to the tiger. The big cats should be kept in the environment that is best suited for them.

starting with the basics

CHAPTER 3

*The only mystery about the cat is why it ever decided
to become a domesticated animal.
Sir Compton Mackenzie*

REMEMBER THE PREDATOR

When you think of a tiger, it's easy to see a predator: the massive
head, three-inch canine teeth, razor-sharp claws, and eyes that reveal
an appetite to devour seventy pounds of meat for one meal. But what
about your cat at home? Of course she doesn't have the belly to hold
seventy pounds of food, but the predator is still a part of her. All cats
are predators. It's as simple as that. The ultimate hunter, graceful,
sleek, and deadly. All cats are of the order Carnivora, meat eaters—
animals that prey on the prolific vegetarians that abound on the earth.
Attacking the weak and old, they are one of the many checks and
balances that nature provides against overpopulation in other species.
They live in all regions of the world, big and small, fulfilling this
ecological role.

These days, your cat's hunting skills may not be overly developed, but the primitive instincts are still there. Think back and remember a look your cat gave you—one very intense with a flicker of
green, making you feel a little unsure. Or a cat fight, with those shrill
screams. The flicker of a cat's tail when she is intently stalking is
hypnotic. As you watch, it's easy to stop breathing as you anticipate

**ALL CATS ARE CARNIVORES AND
HAVE BEEN TRACED BACK TO
SMALL CIVETLIKE ANIMALS
CALLED MIACIDS, WHICH LIVED
MORE THAN 65 MILLION YEARS
AGO IN A WORLD DOMINATED BY
GIANT REPTILES, THE DINOSAURS.**
●

the pounce. These moments are reminders that your friendly companion still has some of her wild ancestor in her. As she is curled up in your lap, softly purring, it's hard to believe that the cat has been domesticated for a shorter time than any other domesticated animal.

To establish a good relationship with your cat you must understand her history. Over the last thirty-five hundred years people have forgotten the cat's past. We have been so taken in by the cat's elegance and charm, her original function has long been forgotten. The rodent catcher is now the windowsill slumberer. Forgotten are her duties in pest control, her role in nature's checks and balances. Now she shares our food, the warmth of our homes, and doesn't have a care in the world. She is the domesticated cat. We trust her completely. She comes and goes at will, left unattended to act as she pleases, and if called, chooses whether or not to answer—a luxury that "man's best friend" has almost never had.

The cat has established a special niche in society, the perfect companion with simple needs—but still problems can arise. If you have ever been scratched or bitten by a cat or have become frustrated trying to understand why your couch has just been shredded, remember the heritage of your companion and the background from which she comes.

Keeping your cat's history in mind, you can begin to condition her wild instincts which all cats possess. For a happy relationship, this aggressive hunter side of your cat must be shaped and modified.

THE SABER-TOOTHED TIGER, WHICH WAS PLENTIFUL IN EUROPE, ASIA, AFRICA, AND NORTH AMERICA ABOUT 35 MILLION YEARS AGO, WAS CAPABLE OF HUNTING A FULL-GROWN ELEPHANT USING HIS MASSIVE DAGGER-LIKE CANINE TEETH.

•

KITTEN CONDITIONING:
TAKING OVER THE MOTHER'S ROLE

When kittens are born they are completely dependent on their mother. Blind and deaf at birth, they look only for food and warmth. They learn about the world around them as they develop, using their mother as the focal point. Everything centers on the mother and she does not disappoint them. A female cat constantly cleans and attends to her kittens. She doesn't share this duty with the male and instinctively her need to care for her young is very strong. Hence a very strong bond develops between the kittens and their mother.

The best way to start conditioning your kitten is for you to take on the mother's role. You need to provide comfort, discipline, love, warmth, and caring for your kitten. By giving these things and spending time with your new kitten in his formative years, you will establish the bond and relationship that will last throughout your cat's life. In the beginning, there are several essential things to do for your kitten:

- If you have the opportunity, watch the mother and kitten in action: Observing the bond they share will help you better to understand the role you are taking on.

- You need to make your new kitten as comfortable as possible. Put him in a room that's warm and quiet without a lot of places to hide in.

- If you have just taken the kitten from his mother, see if you can get a couple of towels that have the mother's scent on them—this can easily be done by gently rubbing the mother cat with the towels. The scent may not be noticeable to you, but to a young kitten it can have a calming affect. If the mother isn't around, use some old clothing of yours as bedding and make it comfortable and as inviting as possible. This will help your kitten get used to your scent.

- A lot of touching in the beginning stages with your kitten is crucial. Kittens need warmth and physical contact—touching—and without their mother they will look to you for this security. Go slowly and let the kitten investigate contact with you as much as possible in the beginning stages. Remember always to use a gentle touch.

- You need to be readily available and give as much attention as possible. Keep yourself in close proximity to your kitten, giving him ample opportunity to come to you.

Through the years, I have been lucky enough to see all phases of a mother tiger's relationship with her cubs. There are few things in life more moving than watching this interaction. The

mother, who is so strong and large, is ever so careful and gentle. You can learn a lot watching the way she sets rules for her youngsters to follow right from their birth. It can be difficult to detect these guidelines as you watch the cubs run around, using Mom as a play toy, just a blur of motion. But even during this time, Mom is setting definite standards for the cubs to follow.

If a youngster is having a little too much fun at Mother's expense, such as biting too hard on the tail or using her as a springboard, she will give him a harsh look. If more discipline is needed, she will reach out and pull the young cub over to her and hold him under her giant paw. She will also take advantage of the time he's immobilized to give him a face bath, and at first the cub will squirm around, trying to escape and wanting instead to play. Her actions are simple and the lesson learned by the cub is: Bite Mom Too Hard, No Play.

The mother teaches her young from the very early stages and the cubs learn to listen. They look to their mother for guidance, come to her for protection when they're frightened, and need her approval when they're brave. When you take over the mother's role for your new kitten, you will need to be just as reassuring and set even stricter guidelines. No biting at all can be allowed, for example.

When you bring home a kitten, you are accepting full responsibility for him. You are the mother and the kitten will look to you for guidance, nurturing, warmth, comfort, and love. What comes back from the kitten will be understanding of your rules, warmth, and devotion.

WHEN IS THE BEST TIME TO START CONDITIONING YOUR KITTEN?

When you get a kitten it's best to start conditioning immediately after you arrive home. Remember it isn't wise to take a kitten younger than eight weeks old as she needs to develop physically and mentally with her mother until that time. By that time, too, all vaccinations should have been started and the kitten should be finished nursing and on solid food. At this stage she will be alert and on the move. If you adopt a kitten older than eight weeks, be sure to find out what kind of human interaction, if any, was given to her by the previous owner or animal shelter. The most formative time of a kitten's life is between three weeks and three months, so early human interaction is best for future behavior development.

If you decide to keep a kitten born from a household cat, at eight weeks it's important to start replacing the mother's time with your own. The more time you spend with your new kitten during the socializing process, the more comfortable your kitten will feel around you. Long periods of quality time with you and without the mother will increase your kitten's receptiveness to you. When your kitten's attention is centered on you, the bonding or imprinting will begin.

At first spend time with both the mother and kitten. Let the kitten grow comfortable with your presence; pick her up, touching and petting. During the day when her mother is up and about, spend time with the kitten, just the two of you. As the kitten becomes more comfortable without her mother (and when the mother has no problem and is not crying out or pacing at the door), separate them for longer periods and use this time to play with the kitten.

* * *

Through the years I have tried many different methods of taking over the mother's role with the tiger cubs. I tried raising them with the mother always there, with the mother there part-time, and finally without the mother after the cubs were four weeks old. In every case, the relationship between trainer and cub was stronger the more hours the trainer spent with the cubs.

When we left the tiger cubs full-time with the mother while trying to condition them, they began to respond only to her discipline. If we tried to do anything but play "cub games," they would be aggressive with us and run to their mom for protection. I have never been nipped at more times in my life than in this circumstance, even though the mother herself had a great temperament, was well trained, and would never show such bad manners!

The second method we tried was letting the cubs spend four hours a day with their mother and trainers and the rest of the time with only the trainers. This method proved to be effective because when the cubs spent the majority of the time with the trainers, they would listen to the rules and standards being set. During the time spent with their mother, the cubs could play their usual rough way and their mother would spend a majority of the time grooming and bathing them with her large tongue. When she was finished, they were always clean, soft, and tired.

The best results were achieved when we took over the mother's role and the cubs were with us full-time. The mother tigers have no problem turning this role over to the trainers

because of the trust they have in us. That the trainer has assumed the mother's role is immediately accepted by the cubs and does not change throughout their lives—the same is true of house cats. And the time and commitment needed to take over every task of a tiger cub's mother is a full-time job.

Taking over the mother's role with your kitten is a full-time job, requiring time and commitment for you to establish a strong bond and relationship with your new kitten. It is your job to teach your new kitten the rules of the house. If you live with the kitten and her mother, you will become an extension of the mother. Mother cats of course have different sets of standards from the ones you will be setting. A cat won't grab her kitten by the ear for tearing apart the plants or climbing up on the couch. But if she does see a possible danger, she will teach her kitten to lie down flat or hide in cover. The best thing you can learn from watching her teaching tactics is that she is consistent! She knows exactly how rough the kittens are allowed to be when they play with her and will reprimand them every time they push beyond the limits she has set. Consistency is the key to establishing rules and conditioning your kitten to follow them.

ABCS OF CONDITIONING

Fundamental techniques of shaping all animal behavior are basically the same. The most important rule we call the ABC of conditioning: Always Be Consistent!

Cats are creatures of habit and are confused easily. The best way not to confuse your cat is to be consistent with your own behavior. For example, if you want your cat to understand the word "no," you must say only *that* word, using the same tone of voice each time. If you get excited or forget and say, "Don't" or "Stop," you will only confuse your cat. Whatever it is that you want him to do, you must first think your method through, and once you start, do not deviate. Be consistent.

Conditioning is much easier with a kitten in his formative years and will take more time and patience with an adult cat. But you can always achieve results when you keep in mind these simple points:

Give Affection: Center your attention on your cat. Touch. Pet. Cuddle. Show warmth and kindness. Lie on the floor and let her rub against you. If you can, let her rub against your face just as she does with her mother. Building this kind of a relationship will make your kitten want to be around you more and make her easier to condition.

THE ANCIENT EGYPTIANS
TRAINED CHEETAHS TO ASSIST
THEM WHILE HUNTING.
●

Make Time: With today's busy schedules, time is not always an available commodity, but it is very important to a young kitten. A disciplining mom will be there twenty-four hours a day during the developing stage. The time you spend doesn't always mean conditioning or giving your kitten attention. It's also important that he feels comfortable playing alone around you, or sleeping by you. For a kitten, life is too interesting to spend every moment with you, but it's nice and comforting to check in with the mother figure and get a loving caress every now and then. The more time the two of you spend together, the better results you will have when molding your feline's behavior.

Patience Makes Perfect: How does that saying go? "Lord, I pray for patience and I want it now!" When you're working with any animal, especially a kitten, patience is a necessary tool for observing and learning to understand that animal's nature. It is also essential in order to reach the goals you set. Because cats are very instinctive and quick to react, it takes a tremendous amount of patience when conditioning one. But if you have the patience, take your time, and are consistent, you will be surprised at the results you can achieve with any cat. Having the patience to go over a task again and again is important. Don't forget that many times a cat will misunderstand what you want or react to some outside influence, like a nearby bird or a dog walking by the window. Have patience and wait for her to calm down. This will give her a chance to listen and understand what you want.

Don't Think So Human: Nineteenth-century authority on cats St. George Mivart said it best: "We cannot, without becoming cats, perfectly understand the cat mind." Remember cats may act human

ALWAYS REMEMBER; YOU CAN
TEACH AN OLD CAT NEW TRICKS!
●

at times, but they're still cats and they think like cats. Put yourself in your cat's place and try to think like a cat. This will make it easier for you to condition his behavior, eliminate problems, and keep him happy. Try getting down on all fours and crawling around the living room.

Use Repetition: Consistently repeating over and over again whatever behavior you are trying to achieve with your cat will accomplish your goal. It may, however, take a considerable amount of time, especially with an older cat. And sometimes you may feel that she will *never* catch on to what you are trying to achieve, but hang in there, it will happen.

Time-outs: Conditioning your cat's behavior can be done only when he is in a state of mind to listen to you. If your cat is too excited or is becoming frustrated with you, then take a time-out. Give him a chance to calm down and refocus his attention. Just stop whatever you are doing. A time-out can last anywhere from three seconds to three minutes—the cat's response to the "time-out" will dictate the duration.

Quit While You're Ahead: Your cat's attention span will tell you how long you can work with her on any kind of conditioning. You need to stop while she is still interested in you and whatever you are working on together. Stopping *before* she gets bored or frustrated will keep her interested and ready for next time.

Establish "No": The word "no" is a simple word that is very important. It can be a great tool or, if not used correctly, a hindrance. In the beginning, the word "no" has little meaning to a kitten. A mother cat conveys "no" either by a cat sound or with body language. A mother tiger roughhouses with her youngsters if they do something wrong, or she holds them down. If they persist, a snarl is her next way of saying a harsh "no!" Rarely will Mom go the next step of nipping or biting at the cubs. When the cubs are young, Mom does not overly reprimand them. She will be stronger with them as they mature and it's time for them to find a home of their own.

We need to make our cats or kittens care about the word "no." "No" is a response to a negative action. It's a simple word that has to be used correctly to get the point across. Used incorrectly, or not at all, discipline will go right down the drain.

First, never scream the word. Cats have good hearing and a normal speaking voice will do the trick. Do put enough tone into it to receive attention, but do it quietly enough not to startle your cat. When we yell the word "no," all it does is make us feel better. It makes your cat want out now. That's the wrong reaction. What you want is a response—the cat to acknowledge you and stop what he's doing.

Good Timing: To be effective you must time your reinforcing or reprimanding response immediately after or during an action. For example, to make the word "no" carry some weight, you have to get the timing right. You must use "no" precisely when the kitten or cat is in an act you want to discourage. A delayed "no" has very little meaning. Don't worry, your cat will give you another chance to say "no" later on—you can count on it.

Give Praise: The secret to a successful "no" is *always* to praise or give positive reinforcement to your cat or kitten when he has stopped misbehaving. Sometimes, we get so caught up in stopping a cat's bad behavior with "no" we forget to praise him for obeying us and stopping. The word "no" will never work unless you consistently follow it up with reenforcing praise. Remember, always praise your cat immediately when he achieves the behavior you desire.

Tara was a good test for our conditioning theory. In the beginning, she started a "nip and run" routine and was very pleased with herself for getting away with it. After one nip on my behind, I became attentive to her sneaky ways and watched for her to do it again as she innocently approached me, acting as if she hadn't a care in the world. The key here is being aware of your cat's ways, even when they seem virtuous. Remember: Don't think like a human.

As she began to approach me from behind, I kept my head facing forward, watching her out of the corner of my eye. Just as she went for a second nibble, I quickly turned to her with a stern "No!" She instantly stopped and gave a chuff. I chuffed back and praised her for not following through with her mischievous act. By my consistently repeating this discipline with Tara's "nip and run" attempts, she soon learned that she could not get away with this behavior and stopped trying.

You can see how this can get very confusing for your cat. Everyone who lives in your house or spends time with your cat will need to be consistent with him. Get everyone together and go over the basics of the ABCs or have everyone read this section in the book. Having consistent behavior from everyone in your home will make it much easier for your cat to understand the rules and for you all to live more harmoniously under one roof.

GET A GRIP ON TEETH AND CLAWS

Two good reasons to condition the behavior of your cat are her teeth and claws. Yes, those eighteen razor-sharp claws that are so expertly used to grab prey. Those thirty piercing teeth that can pull apart meat and slice it into bite-size pieces. Cats have the natural ability to use these "tools" with agility and accuracy. When using their claws and teeth, cats can get into an instinctual state of peaked excitement. At this point, it's no wonder that a couch is no match for an unconditioned attacking cat.

UNLIKE OUR MOLARS, A CAT'S MOLARS CAN CUT MEAT EXACTLY AS SCISSORS WOULD.

●

> Working with tigers (whose claws and canine teeth of course are much larger than those of their domesticated relatives) taught me quickly that *all* cats can learn not to bite or scratch. Half the battle is knowing that you can condition against and even change this bad behavior. The other half is persistence.
>
> All the tigers I work with have been taught, beginning in cubhood and continuing throughout their lives, not to bite or scratch. The degree of control I use varies with the situation, such as play, peaked excitement, and outside stimuli. Without this training I would not step foot near a full-grown tiger. But because of the consistent training and conditioning, I feel confident being around the tigers in all situations.

No matter what the circumstance is, our tigers know that a trainer is not a pin cushion, and you need to teach your kitten or cat the same thing. Biting is never acceptable. When your kitten is so small, soft, and cuddly, it's easy to forget she will get bigger and stronger as an adult. Once your cat is grown, she is big enough to cause injury. Kittens love to mouth and bite everything. When they

investigate something, it goes right in the mouth. Pick it up, move it, toss it up in the air. Make it fun, shake it. Experience it. The more fun it is the more exciting it is. When a kitten is in this natural stage of behavior, it's the perfect time to teach her not to bite—not everything, just not you, your body parts, hands, feet, or all the areas in between.

TEETH FACTS
FOR TIGERS AND CATS

Milk Teeth are temporary teeth that start coming in 2 weeks after birth and should be completely grown in at 8 weeks.

Permanent Teeth start coming in at 4 to 5 months and should be completely grown in at 8 months.

There are 30 permanent teeth (12 fewer than a dog has), 16 in the upper jaw and 14 in the lower jaw:

- 4 Canine teeth (dagger-shaped, used for killing prey and tearing)
- 14 Molars and premolars (used like scissors for slicing and cutting)
- 12 Incisors (used with the canines for pulling and holding on)

We've found just how important it is to control biting while the tigers are still cubs. Cubs love to bite and bite hard. A bite feels like a hard pinch and can even leave black-and-blue marks. Kittens that are allowed to bite when young are much more prone to biting when they're older. To get them to stop, we use our conditioning technique and are very repetitive. Here's what we do with the tiger cubs and what you need to do with your kitten or cat:

- To control biting, start as early as possible. Young cats naturally bite, adults cats naturally bite harder.
- Every time your cat bites, look at her and say "no." If she does not stop, either tap her on the nose or place your hand on her head and hold tight (which is similar to the mother tiger's immobilizing with her paw). When she stops biting, praise her in a gentle voice.
- When a cat is excited and appears about to bite, stop all movement. Any kind of movement will make her more excited.
- Because of a kitten's natural tendency to mouth things, she will give you plenty of opportunities to teach her not to bite you. Use these opportunities consistently.
- You can control outside stimuli that may excite your cat into biting by being aware of and avoiding them.
- You too must maintain control and concentrate on conditioning good behavior—it makes a difference. Some people get angry if bitten: "How could she do that? I feed her and take care of her." Some of us lose control, yelling things that can't be printed here. By that time the cat is long gone, along with the chance to control biting behavior.

CATCLAWS

There are 5 claws on each forefoot. The first one, the dewclaw, is set higher than the rest. There are 4 claws on each hind foot. The claws are normally retracted into protective sheaths. This allows a cat to move quietly and keeps the claws sharply pointed, not blunt like a dog's.

We don't see them much, but a cat has eighteen sharp claws sheathed within his soft, furry toes. If you've ever been scratched, then you know why you want to control a cat's claws. It can be very startling and painful—one of the things that people find most offensive about cats.

People often ask, "Don't the tigers scratch you, rip your clothes, rake your arms...?" Occasionally we get an accidental scratch, but all of the tigers have gone through basic training. They know not to scratch and actually think about keeping their claws retracted in our presence.

Most of the time cats' claws are in—it would be very uncomfortable for them to walk around with them out. Cats use their claws for a variety of reasons, but it is usually to pull something in, or to grab what is exciting them—perhaps a toy or small prey. What should not be exciting to them is any human body part. If they claw, it's because they are out of control. They are excited and are just reacting.

To stop clawing, you need to start by making your cat think about his claws and controlling them even when he's in an excitable state. That's how we control the tiger's claws. They learn their claws should not be out in our presence and if they scratch, it's because of an accidental reaction, not because they went out of their way to be aggressive.

Kittens and tiger cubs become excited at anything that moves. They bat it around and become even more excited. The claws come out to pull in the object, to roll it around. This natural behavior looks cute, and often we don't take offense at it even when they give a small scratch. But do not ignore or forget this little scratch because when a kitten grows into adulthood, his claws get bigger, more efficient, and more capable of damage.

An unconditioned tiger cub will scratch a trainer up and down, anywhere his paws can find. Your kitten will have the same tendencies if he is allowed. You need to begin as soon as possible to teach your kitten to keep his claws in. The method we use is effective and works as well with an adult cat, only it takes more time.

- The first time your kitten or cat extends his claws and a paw makes contact with you, or you receive a little scratch, during play is the time to begin his training. It's up to you to control his claws from this first moment on.

- When the claws come out, look your cat in the eyes and say ''no.'' If the claws don't go in, tap the paw, and say ''no'' again. Cats, kittens, and tigers don't like to have their paws touched and will almost always pull their claws in when their paws are tapped. If your cat is being stubborn, then lightly pinch his toe while repeating ''no.''

- Once you start, you have to be consistent. Your kitten is looking to you for guidance; you are the mother figure. The discipline is up to you every time you see his claws come out.

THE CHEETAH IS THE ONLY CAT THAT ALWAYS HAS HIS CLAWS OUT. THEY, LIKE THE TOENAILS OF A DOG, ARE NOT RETRACTABLE.

●

- If a cat scratches and runs, it won't do any good to chase him around. We've tried it—and it doesn't work. Cats forget more with every foot they run. Wait until the next time.

Training a cat about his claws is best done during calmer moments when he is playful but still in control. The middle of a high-energy play session is not the time. Your cat will test you, acting in character and out of character; you need to have the patience to find out what is going on in his head. And by your controlling his behavior consistently, he will soon think about keeping his claws in around you. It's worth the time spent, and you won't have to worry about being bitten or scratched.

WHY THE *@%* DID HE DO THAT? REASONS FOR UNEXPECTED BEHAVIOR

Once you understand and use these conditioning methods, if your cat reacts in a manner that you feel is aggressive, such as unexpected biting or scratching, consider that there may be a reason and try to find the cause of the behavior. Then, you will be able to understand and correct it. Some common causes for unexpected aggressive behavior:

- **A Sore or Injury** If you suspect that a sore or injury may be causing excitability or aggression, wait for your cat to calm down, then examine his body thoroughly for any indication of harm.
- **An Outside Stimulus** It could be that another cat or dog, a neighbor, or something else has frightened your cat. She may have transferred this fright to excitement and taken it out on you. Look around to see if anything out of the ordinary has happened to cause

this reaction. As you spend quality time with your cat and understand each other more, you will be able to predict when outside stimuli can cause aggressive behavior and you will be able to detour any aggressive reactions.

- **Lack of Activity**   If your cat has been closed up in a small area, or is accustomed to roaming freely and has not been able to, aggressive behavior may be the result of pent-up energy being released. Even in new situations, be sure that your cat gets ample daily exercise. Not only does this help him stay calmer with you, it helps keep him fit and healthy.

- **Hormonal Change**   If your female is not spayed, her behavior will change when she comes into heat. If your male is not neutered, his behavior will change when he is around a female in heat. The female becomes more vocal and much louder. Both sexes can become antsy and frustrated. Watch carefully during this time, since this is when aggressive behavior can occur.

I once had an experience that is a perfect example of what can happen as a result of an outside stimulus. One day Mohan, the ultimate majestic-looking male Bengal tiger who weighs in at a strong five hundred pounds, came out on to the grassy area of Tiger Island with me late in the day. Baghdad was in heat and Mohan knew it. She and the other females were on birth control pills at the time, which make her react "more distant" toward Mohan than when she is not on birth control. Mohan thought the time was right for mating, but Baghdad would have nothing to do with him. This made him extremely frustrated. To make matters worse, one of the male tigers whom Mohan is not particularly fond of had been on the grassy area earlier and had left numerous "calling cards" all around by marking his scent everywhere.

Not being aware of what a tough day Mohan was having, I went over to the area where he was sitting to join him in some quiet time. What happened next was anything but quiet. As I started to sit, Mohan leapt up and bit me on the right shoulder. I fell to the ground with his two enormous paws on my chest. He stopped and looked down at me with a look of excitement and confusion on his face. He knew it was wrong to bite, but the outside stimuli had frustrated him to the point of no control. I gave him a stern "no" to make him focus and regain control. He immediately realized his wrongdoing and ran off.

This is a typical example of displaced aggression caused by an outside influence. Mohan bit me because of frustrations not having anything to do with me. I did not take his actions personally at all, but rather blamed myself for not being aware and noticing the circumstances that made his behavior change.

Always look for any outside or hidden reason for your cat's unusual behavior. The more time you spend with your cat and the more the two of you understand each other, the easier it will be to understand and predict "quick reactive" behavior.

playtime with a furry friend

CHAPTER 4

*There is nothing in the animal world, to my mind,
more delightful than grown cats at play. They are so
swift and light and graceful, so subtle and designing,
and yet so richly comic.*
Monica Edwards

WHY IS PLAYTIME IMPORTANT?

The golden eyes are wide with excitement, the striped body
tense and low to the ground. I pull the rope little by little, then
stop. Kimra inches closer, her body rock-hard and ready to
pounce. Some inner voice tells her the time is right and she
springs. She races toward the innocent looking object . . . the
instant before she reaches it, I yank the rope high into the air
and Kimra follows, jumping ten feet straight up. In midflight
she reaches high and grabs the rope. In one fluid motion she
pulls the rope into her mouth and lands on the ground.
Nothing can escape the great predator.

Kimra is an adolescent tiger, a 150-pound spring ready
to uncoil. The game we're playing is "chase the prey." In many
ways it's like playing with cats at home. It looks like an
innocent game. But it takes months of conditioning to keep the
tiger in control during the excitement of playtime.

73

We love watching cats in action—the spring, the charge, the wild excitement. It's thrilling and many times funny to watch their antics, but in order to avoid accidents both tigers and cats need to be controlled by certain limits during playtime.

One of Bill's major goals in the last ten years of working with the big cats in natural habitats is to keep them active and stimulated. Whether by coaxing their natural swimming behavior from them or enticing them to "hunt" for exciting toys, he gets the tigers to release as much energy as possible. This playing keeps the tigers much more interested, active, and healthy. By relieving boredom for your cat, you are creating a less stressful environment. Animals who live in stress-free and stimulating environments lead longer and healthier lives.

All cats need play for activity and it is up to us to make sure they get their daily requirements. When you have an indoor cat, it is entirely your responsibility to supply activity. This activity and play are very important for your cat's physical and mental well-being. But if playtime is not properly conditioned, it leads to "pain time."

A cat in play spends a great deal of time practicing his hunting skills. Every cat is born with the instinct to hunt, but his skills need to be developed. A bug, leaf, butterfly, fluff ball, socks—you name it—all these can be the object of your cat's mock hunt. It looks playful and amusing to us, but the cat is honing his most basic instincts.

As you're playing and having fun with your kitten, knowing that there's a good chance that the two of you have different ideas in mind will help you set consistent standards for playtime, which will make playtime more fun for the both of you.

PLAYTIME ETIQUETTE

We once helped a family whose tomcat was allowed too much rough play—a classic example of an unconditioned cat. This eighteen-pound five-year-old named Turbo had his own playtime routine: Jump up and grab onto your hand with his paws. Because he was allowed to do this since he was a kitten, a hand soon became a toy or the prey that really got Turbo excited. You could not pet him without his latching onto your hand, and of course all dangling hands from innocent bystanders were fair game. Many people were scratched and bitten when Turbo got particularly excited. This play conduct was allowed by one member of his family and quickly became a bad habit for Turbo. Changing Turbo's behavior by getting him to play on our terms took a long time. Turbo soon learned to enjoy playtime without pain to his partner. Situations like these can easily be avoided when playful ways are conditioned from the very beginning.

Controlled Excitement: A playing cat is an excited cat who can get out of control—it's her nature. It is up to you to control the

level of excitement. During the tigers' playtime Bill is always aware of their excited state. If you are not constantly alert to your cat's state, you may suddenly be faced with a cat acting purely on instinct and not listening to you. This is not an option with a three hundred-pound tiger if you want to stay alive. The signs of an excited cat to watch out for: Body stance becomes stiff, the head moves from side to side, the eyes dilate with a wild look (when extremely excited, the eyes show a flicker of green). Usually the best way to judge is by your gut feelings. If you feel your cat is getting too excited, take a time-out. Stop all activity until she has calmed down enough to the state where she will listen to you. Then praise the fact that she has calmed down and return to playtime.

Walking Away: If your cat suddenly gets very excited during play, the last thing you want to do is reach out and try to soothe her with a pat. This can trigger a quick reaction you both will regret. When you've tried a short time-out but she is still very excited, *walk* *away* or at least put some distance between the two of you. This gives your cat an opportunity to calm down and is your way of showing that you stop play altogether if she doesn't control herself.

Hands Are Not Toys: Even when your cat is a tiny kitten, *never* let him play with your hands, which is one of the most common mistakes cat owners make. When playtime gets a little carried away, this bad habit will always turn into a bite. It is extremely important not to let your kitten or cat use your hand as a toy *ever*, or you will confuse him by giving him mixed signals. If you look at a hand from your cat's perspective, you too would get excited by this small object with five individual moving parts! Although "wiggly finger" games can appear fun when your cat is a kitten, they create many bad habits later on. Keep playtime etiquette simple and consistent from the start by never allowing your hand to be a toy.

Toys: Since your hands aren't to be used during play, toys are the best alternative to use during playtime with your cat. Toys give your cat a chance to play in true form, when using her teeth and claws is permissible. To a cat, the whole world can be a toy, depending on her mood, of course, and whether the toy is wrapped right. Be creative when looking for toys for your cat. Both of you will have fun and you can even make your own (see pages 84–85).

Playtime Routine: By using toys, you will be able to develop a playtime routine that will be easy for your cat to understand. When your cat sees a toy, it's time to play. If there are no toys out, then he knows to find something else interesting to do or enjoy quiet time.

Establishing *Your* Definition of Playtime: It's important from the beginning for your cat to understand what playtime is and that she must control her behavior before her excitement turns into aggression. Start slowly to be sure that she understands this. A kitten will easily get excited, so you need to take frequent time-outs. Praise her every time she settles down. Soon she will understand that playtime is fun, but also that she must stay in control.

You can control active playtime to fit your own schedule. Playtime is a great time for your cat to exercise and get rid of excess energy. You can schedule playtime so that all her energy isn't being expended at night when you're trying to sleep or at other inappropriate times.

WHERE TO PLAY

Where you decide to have playtime with your cat is important for both the safety of your cat and your household breakables and furniture. Once you decide where this play area is going to be, be consistent. Cats have a difficult time understanding that it's okay to play here one time but not the next. Choose an area in which you feel secure, where there is nothing sharp to injure your cat and where you can live with any damage to objects that occurs. Remember, this area is just for playtime. You need to show that playtime happens only in places you allow. (Rakon, one of our largest male tigers, understood this rule so well Bill once brought him into a crystal store for an appearance!)

Never play on or around the furniture. You don't want your cat reenacting unsupervised play later at the expense of your furniture. Overstuffed chairs and couches are great for quick turns and claws itching to get into something. Help your cat out by consistently staying away from furniture during playtime. Watch out for play areas near corners in rooms with breakables. What about when you're not there and your cat wanders in looking for something to do? That could end up with your cat hurt or one of your favorite knickknacks broken after a playful wrestling match. Cats are creatures of habit, and if allowed to play all through the house, accidents can happen. Be consistent and keep important rooms off limits during playtime.

Outside areas are a great choice for playtime, depending on your location and climate. It's not only much easier on your belongings but a good opportunity for the two of you to get some fresh air and sunshine together.

RELIEVING BOREDOM

To stimulate all of your cat's keen senses, you need to be creative during playtime. Getting the most out of playtime will depend on what toys you choose and how you use them. Your cat's best sense is her sight, her second, hearing. Smell is her least reliable sense, but it is still much better than that of humans. To keep your cat at high levels of activity throughout her entire life, and to prevent boredom, you need to keep her interested. Stimulate all of her senses. Here are some ways:

Colors: It used to be thought that cats were color-blind, but this has been disproved in recent years. As you know, objects that are small and quick look like prey and are always sure to get a cat's attention. So try cat toys in different and bright colors to stimulate your cat's sight and entertain him for hours. Pet shops now carry many different-colored fun toys such as assorted colors and shades of balls and catnip bags in different colors. Or there are "inexpensive" toys you can provide yourself such as different-colored paper crinkled up into balls, different-colored paper bags your cat can lose himself in, a variety of colored ribbons, and so on. Too, you can experiment with your own ideas.

Tigers are very influenced by color changes. We regularly use this to our advantage in tiger activity. Changing the color of the rope attached to an old toy makes it a brand-new toy to a tiger! If a tiger is lethargic on a hot summer day, we change our shirts to ones of a different color and instantly get all the tigers up and running. After playtime, we put on the normal-colored uniform they are used to seeing which signifies that playtime is over.

Movement: During playtime, all movement stimulates your cat, but cats are much more alert to sideways than up-and-down

movement. Have you ever dropped an object straight down in front of your cat's face—and she completely misses the movement and keeps looking at you, ready to receive one of her favorite toys she saw in your hand just moments before?

Rakon's favorite thing in the world is food. More than once I've had a tasty treat in my hand to give him and accidentally dropped it straight down to the ground. I always chuckle as he continues to look at me for his reward, because as a rule Rakon never misses a trick when it comes to food! If I throw it horizontally, he's always on it in a flash.

During playtime, make your toys even more interesting by using horizontal movement. Throw a small toy from one side to another low to the ground, slowly so your cat can follow it. This movement simulates fleeing prey and will get your cat really excited.

Partial Vision: Since cats' sight is their strongest sense, toys and playtime can turn into loads of fun by hiding or partially hiding an object. This works particularly well if your cat loses interest in a particular toy or game. For a cat, anything that is partially hidden is much more fun than if it's right out in the open. In the wild, cats use cover, staying low to the ground, hiding behind vegetation, as they slowly creep up on their prey. Because of this low angle, they usually only partially see their prey.

For Kimra, if a toy is out in the open, it's no fun. But put one of her favorite toys behind a tree or large rock where it's partially blocked from view, then it's a blast! As she sneaks up on the toy, I like to move it slowly to really get her attention.

Try hiding one of your cat's favorite toys, and when he spots it, slowly move it with an attached string. You'll find that hiding fun things and giving your cat only a partial view of them opens up a whole new avenue to playtime.

Sound: Another way to stimulate your cat during playtime is with sound. A soft or low sound can be enticing to a cat. Your cat knows that nothing moves in silence and the sound of slow movement will definitely get her attention. Attach a toy to a stick or pole so you can make a scraping sound across the ground, or tap it upside down to make sounds similar to scurrying prey. Do anything you can think of to get your cat's sense of hearing involved. The more creative you are with sounds, the longer you will be able to keep her interest.

Smell: Cats tend to use their sense of smell as a last measure, but when they do, it can be very stimulating for them and get them excited. You can use catnip and rub it onto any toy to entice your cat's sense of smell. Tying a string onto a bag of catnip and hiding it behind something stimulates all of your cat's senses! If your cat is the type that gets overly excited with catnip and starts to get out of control, just use small amounts at a time. What works best is rubbing the outside of a soft toy with catnip. This will get his attention, and the catnip wears off quite quickly and doesn't overexcite.

I have tried catnip with the tigers at all different ages and found for the most part they were very indifferent to the substance that can really get their smaller cousins excited. For some reason, Baghdad was afraid of catnip, while the others would sniff it and just walk away. Rakon, being the character he is, found his own alternative to catnip—a juniper bush. He became extremely attached to the bush and would roll around on it getting extremely excited, just like a cat with catnip. This wasn't too good for the juniper bush.

Stop While It's Still Fun: Always stop a game while your cat is still interested. This will make her much more attentive and eager for next time, keeping playtime a positive activity for her.

Fridge, our tomcat, put us on to "stopping while you're still having fun." His favorite toy turned out to be a fly swatter that was innocently hanging from the top of the pantry door. During the summer months he developed a game that would entertain him for hours. He would start out by staring at the swatter, pacing back and forth, anticipating his attack on it. Then he started jumping at it and was soon jumping the full six feet to knock it from the pantry door. Once he achieved his goal and his favorite toy was on the ground, it did not have the same appeal. To keep him interested in this game we invented a playtime routine. We put the swatter in a drawer and brought it out only occasionally. As soon as Fridge saw the swatter, he knew it was time for high jumping and got very excited. If he knocked it down, I would put it back and the game would continue. Before Fridge could lose interest in this game, I would put the swatter back in the drawer, letting Fridge know playtime was over. Exhausted from springing into the air, Fridge usually would turn in for a long nap (probably dreaming of fly-swatter-shaped prey).

Playtime with your cat is the perfect opportunity for you to leave your worries behind. Putting out her favorite toy and watching her in a playful fury is a stimulating way to relieve your own stress. Laughing is an extremely beneficial therapy we all can enjoy while watching the comical acrobatics of a cat at play.

TERRIFIC TOYS

Anything can become a cat toy. Toys for your cat can be store-bought, handmade, natural, or anything you and your cat can come up with. Here are some suggestions for homemade toys:

- Anything tied onto the end of a string attached to a stick is an irresistible toy for your cat. Attach a toy, ball, or piece of cloth to a heavy-gauge line fastened to a fiberglass pole. You can either dangle it in front of your cat and watch him leap and do somersaults for it, or you can move it across the floor and watch him stalk it.

- Hanging one of his favorite toys from a stationary point by an elastic or rubber band will also get any cat going in an energetic fury of play.
- An old empty brown paper bag laid on its side is a great toy that will keep your cat entertained for hours.
- Ping-Pong balls are always a sure way to really get your cat going. Roll one toward him on the floor or try several to really get him excited.
- An aluminum foil ball is entertaining to your cat because it's shiny and makes noise as it's flung across the floor. When you use the foil balls, always supervise the playing and be sure you take them away when playtime is over. Remember, putting your cat's toys away while he's still interested in them will keep them intriguing for next time.

SAFE TOYS

Always make sure that the toys you or your cat have picked will not harm her in any way. All new toys should be introduced to your cat with your supervision. Even an innocent-looking ribbon can be swallowed.

Check for any small or loose pieces that can be swallowed.

Check for any sharp or rough edges that could cut your cat's mouth or paws.

Be sure the toy does not have any harmful chemicals or paints.

Because of a tiger's strength and possessive nature, we need to be very careful whenever we introduce a new toy to each individual tiger, keeping tight control. We've found that one of the best tiger-proof toys, and one of their favorites, is a beer keg (empty of course). They can't bite it and destroy it within seconds; they can't pick it up and run away with it or hide it somewhere. It's great.

We attach a chain to it and throw it in the pool. Sampson always jumps right in after it, madly pawing away. This toy was great for teaching young Sampson that he can't keep everything he gets his paws on—he was very possessive with other toys. When a cat gets possessive he will snarl, then growl, and get so excited that he forgets all training and returns to instinctive behavior. Well, Sampson didn't quite know what to think when he was introduced to the strange-looking keg. He kept climbing on top of it and the keg would slowly roll over, dunking Sampson underwater. Sometimes it was hard to tell who was winning.

All cats large and small can become possessive of toys or food. It's a natural response for them to latch onto something and sink their claws and teeth into it. It's always easy to get your cat up and chasing a fun object, even when he's been lounging, but it's not always easy to get him to give the object back. Have you ever reached over to take a toy away from your cat and was startled by a stiff growl? That's possessiveness! Many times your feline will not show this unless he is very excited, or has a good appetite and you're messing with his meal.

To teach Sampson not to become possessive over this toy, I started repeating the command "Leave it" as I slowly pulled the keg away. Sampson's first response was, of course, a menacing snarl. I would take the keg away and give him a short time-out until he changed back to his carefree self. Through much repetition, Sam finally learned control with all toys.

This works just as well with your cat. You need to take a short time-out to give him a chance to calm down, then start again in a soft tone, repeating to your cat "Leave it" until he lets you have what you want and you praise him for a job well done. As always, be consistent when managing your cat's possessiveness of anything.

A FIT CAT

Finding fun toys for your cat and setting a regular playtime routine with lots of activity will help your cat have a long, fit, and healthy life. Cats have a reputation for being lazy and not needing regular exercise, but your cat *does* need activity and stimulation. And the best way to achieve this is through playtime.

In the wild, a cat stays fit through hunting. Playtime lets your cat practice her hunting skills and this keeps her muscles and heart strong.

Active play also plays an important role in the mental well-being of your cat. Boredom and minimal stimulation can result in high stress levels and cause many health problems. Excessive pent-up energy will also lead to aggressive behavior.

You need to give your cat a fun reason not to sleep all day. Your playtime routine should differ as well. If your cat were in the wild, she would not be stalking prey in the same place and the same way every day. Have you seen the look of a tiger in a cage, expressionless, as if he were looking right through you? That's because day in and day out, his routine stays the same and he has no stimulation.

It's easy to establish a daily playtime routine that will expend considerable amounts of cat energy and relieve stress and monotony. A fit cat is a happy cat. You will be able to see this in her eyes as she comes over to you, exhausted from a lively play session, laying her head gently next to you, softly purring away.

cat naps

Are cats lazy?
Well, more power to them if they are.
Which one of us has not entertained the dream
of doing just as he likes,
when and how he likes,
and as much as he likes?
Fernand Mery

WHY DOES MY CAT TAKE SIXTEEN-HOUR NAPS?

We've all had a laugh over the strange places and positions in which we've found cats slumbering. Although their reputation for being the laziest creatures in the world is unjustified, still, slumber time is very important to a tiger and cat alike.

Rakon is an energy-efficient cat, slow in movement and quick to relax. Compared to Kali and Serang he appears sluggish and clumsy. During play, they both run circles around him. Often, it's easier for him just to lie down rather than keep falling farther behind. He has perfected and refined a cat's natural ability to conserve energy, with never a wasted step or movement. Rakon goes into the deepest sleep of all the tigers. He will not awaken unless there is an important reason—like food. His ears always perk up at the sound of the refrigerator

door opening, for that's where the meat is kept in cold storage. He lifts his head, sniffing into the air with his large snout, still not sure if it's worth the gamble of getting up. Once he is certain it is a tasty morsel he smells, he's up and running.

Seeing Rakon in this state, knowing well that food is on his mind, I usually ask him to do something simple like stand on my shoulders to show his great size. He slowly obliges, eagerly awaiting his reward. Once the treat is devoured, he trudges back to a shady spot to relax again. I am always amazed to be able to go over to Rakon when he is fast asleep and pick up his head, waiting for a response. His enormous head just drops with gravity when released, with absolutely no reaction whatsoever. But when he hears that refrigerator door, watch out!

SEVENTY PERCENT OF THE TIME A CAT SLEEPS, HE IS IN A LIGHT SLUMBER. THE REMAINDER OF THE TIME HE IS CATEGORIZED AS BEING IN A "DEEP SLEEP," THE KIND WE SLEEP THE MAJORITY OF THE NIGHT.

•

Why do so many members of the cat family need so much rest time? From any cat's point of view, it is not from laziness. Cats prefer getting their sleep in a series of naps—catnaps. If you add up all this nap time, it comes to about sixteen hours a day. But you need to understand what your cat is really doing when he's sleeping. A catnap is a light sleep that allows the cat to restore energy while at the same time staying alert to his surroundings.

Cats have a great sleeping system. Because they are basically solitary animals and need to be alert at all times, a light sleeping pattern allows cats to receive sensory messages as they sleep, which gives them added protection against danger. Many cats filter out most noise and respond only to the sounds that interest them (just like Rakon with the refrigerator). You've probably seen times when your cat has suddenly picked up his head and looked around, then his expression has gone blank and he has gone right back to sleep. Have you ever wondered what it was that initially caught his attention? Only your cat knows for sure. There have also probably been times when your cat has misled you, seeming to be in a deep sleep. You'd be surprised how much your cat is really aware of during this time. French writer François Auguste-René de Chateaubriand observed, "The cat pretends to sleep the better to see . . . "

When you think that your cat is truly in a deep sleep, be careful that you do not startle him! Not only is it a poor way to wake him up, but it could give him reason for a quick aggressive reaction toward you. This would not be an intentional reaction, rather an instinctual one to protect himself. While sharing quiet times together, respect your cat's deep sleep time by being quiet around him.

I learned this lesson quickly from spunky Kali, one of the eight-year-old female tigers. I slowly walked over to her while she was sleeping in the sun. She has the amusing habit of hanging her tongue halfway out of her mouth when she's sleeping. This makes her looks so adorable, it's easy to forget the mischief she gets into. On this particular day, when she looked irresistible with her tongue hanging out, sleeping so peacefully, I decided to give her a tender pat on her back. *Mistake!* In a flash her claws were out and she was all teeth, ready to protect herself. After a few heart-pounding moments for both of us, she calmed down to the state where she "remembered" me. She gave me a very irritated look and went over to find another spot, this time keeping her eyes open.

Cats nap so many hours precisely because they spend the majority of their sleep time in a light sleep. They take short naps all through the day and night because when they are active they need high levels of energy for short periods. Cats are not designed for endurance; they have to be able to have the quick bursts of energy needed for hunting.

Tigers and cats hunt with a slow approach, crouching down low to the ground, inching forward in fluid motion, waiting for the perfect moment. Then they suddenly pounce in the hope of capturing the prey by surprise. This hunting technique uses up the energy conserved and restored by the cats during their naps. You can change your cat's sleeping patterns by using play sessions to keep him active. Also he may need to sleep less if he feels completely safe and secure, and thus be able fall into longer periods of deep sleep.

SHARING QUIET TIMES

Since quiet time is very important to the quality of your cat's life and can encompass up to two thirds of it, sharing it together is important and there are may benefits in establishing shared quiet time. Quiet time gives you the opportunity to strengthen the bond between the two of you, so increasingly your cat will seek out your company during this time. It reinforces your role as mother, as your cat will look to you for warmth, security, and protection. Spending this important part of your cat's life with her counteracts her tendency toward aloof or loner behavior. And quiet time itself provides relaxing and stress-free moments for you to share with your cat.

One of the biggest differences between cats and dogs is that dogs are social by nature. They seek out their owner for attention— put your hand out and the dog is there in a flash whether or not you want him. Cats are not like this, but through quiet time, they can learn to show affection and the ability to please, and not be the loners we make them out to be.

Many times you'll find your cat asleep on top of the refrigerator, in an open drawer, in a laundry basket, and so forth. If you want your cat closer to you and you want to share quiet times together, you need to make her siesta area near you. If your cat never naps in areas where you are, think about your household environment. Is the TV always on or the radio blaring? Is there a lot of traffic running through your house, with family or roommates constantly racing in and out? Do you spend a lot of your time in a cool or drafty room? Are you always in a rush?

To share quiet times with your cat, turn the television off, play some soft music, and sit down and relax yourself. Find a warm spot near a sunny window or fireplace. Your own warm body will be very inviting for your cat to snuggle up to. Cats look for quiet, warmth, and security. It's up to you to create an inviting environment for your cat to share restful times with you.

And you do need to set aside enough time for both play and relaxation with your cat. Relaxation can't be rushed. Setting a soothing and secure atmosphere for her will make her more than willing to get in the habit of sharing her favorite catnap spot with you—if you don't run away as soon as she settles in.

CAT TALK

Quiet time is a great time to let your cat "talk" to you and for you to talk to your cat. But remember one of the quiet time rules: When talking to your cat, make sure you are not interfering with his slumber time.

Cat talk and communication come in a great variety of sounds —from a gentle purr to a hardy meow. With tigers it's a chuff and a great variety of moans and groans. All these sounds definitely have meaning, but only a cat can truly understand them. The same is obviously true for cats listening to the spoken word. They might give the impression of understanding, but usually only a few words are understood and in most cases the words are associated with food.

Most cat owners have great stories about their cat's understanding. Some owners even believe that their cat understands each and every word they say. Cats are smart at being cats, but when it comes to English, they struggle. Your cat's vocabulary is based on sound association. Your cat can learn words such as "no," "sit," "chair," and "jump," but it is a learned behavior and will take time. The best results come when a reward is offered. Think of the sound a refrigerator makes, then imagine that it's a real English word. Your cat knows that word, simple.

So what is happening when we look into our cat's eyes and chatter about the goings-on of our day? Our cat definitely cares and we have his attention.

Often during quiet time with Mohan, the tiger, I will be carrying on a conversation. Mohan is very attentive and even nods at the right times, looking as if he's taking in every word.

THE AVERAGE CAT CAN UNDERSTAND MAYBE 25 TO 50 WORDS, WHILE THE DOG CAN UNDERSTAND A GREAT DEAL MORE.

●

It makes me wonder if cats really can understand—until Mohan yawns right in my face and walks away. A cat is a cat.

Talking to your cat is important not because he hangs on to your every word, but because you are the mother figure. It's natural for a mother to communicate with her young. Even at the first moment of life a young kitten is reassured by his mother's purr. When we talk to our cats, it's primarily in a soft voice that can be similar to a mother's purr. So when they are sitting with us and listening it's not important if they understand; it gives them great comfort just to hear the gentle sound of warmth and comfort coming from their "mother." Remember, talk in a soft voice—and you can say whatever you want. Similarly, you cannot understand every sound your cat makes, but you can learn to sense your cat's meaning from his tone and body language.

A RECENT GALLUP POLL FOUND THAT 9 OUT OF 10 PET OWNERS TALK TO THEIR PETS; 62 PERCENT OF PET OWNERS GIVE THEIR PETS CHRISTMAS GIFTS; 32 PERCENT LET PETS SLEEP IN THEIR BEDS; 30 PERCENT LEAVE THE TELEVISION ON FOR ANIMALS TO WATCH; AND 17 PERCENT KEEP PHOTOS OF THEIR PETS IN THEIR WALLETS.

●

BUILDING CLOSER BONDS WITH TOUCH

Quiet time is also the best time for you to become closer to your cat with tactile communication. Touch is especially important to kittens and comes naturally—they rub against you the way they do with their mother when they want to be cleaned. A mother cat will spend countless hours cleaning and grooming her kittens with her soft tongue, offering a sense of security and protection. The gentle motion of a hand going back and forth is similar to a mother cat's attention. As you are in the mother role now, you need to continue the tactile process throughout your cat's life, building a very strong bond between the two of you.

With adult cats you need to use a softer touch. They need more sleep and rest than kittens, so watch that your soft rubbing and petting is not interfering with your pet's sleep. Many times just a gentle hand resting on her or simply your company and the warmth of your body next to hers is enough when sharing quiet times. You want to be sure your cat is comfortable and enjoying quiet time with you, and that you are not interfering or startling her. So,

- Always use a slow and gentle touch.
- Feel if your cat's body posture is relaxed and her body is not tense.
- Resist the temptation to pick your cat up or constantly talk to her, possibly disturbing her slumber time.
- Touch your cat slowly and gently to give her time to react to your attention, and give you affection back.

Follow these simple techniques during quiet time and your cat will seek you out much more during the day. The best part about quiet time is that your cat is enjoying your company, getting her much-needed rest, and it's not demanding of you!

HOW TO MAKE A COZY CAT

Not all cats are "lap cats" and some may actually resist your touch. Some cats are satisfied simply by being in your presence and do not want to be touched. If you feel your cat isn't as affectionate as you would like, you need to spend time around him and make it as comfortable as possible for him to be close to you. To condition your cat to be cozier toward you, try this:

- At first, start slowly, with some distance, giving your cat a chance to feel comfortable resting with you near.

- Then sit closer to him, still without touching, until you see he is comfortable with you near.
- Once he's used to your presence, little by little proceed to get him used to your touch by starting with your leg or side next to him.
- Start petting him slowly, seeing if he is enjoying the touch.
- Give him a chance to respond.

I went through these steps when first establishing quiet time with Baghdad. At first she wanted nothing to do with me; she would snarl or just get up and leave. But I started sitting near her, with some distance separating us, so she became used to

having me around when she was resting. After much time had passed and the irritated looks were gone, I knew she felt comfortable with me at a distance during her quiet times. Little by little I got closer to her. Sometimes, for each step closer I had to take a snarl and two steps back, but patience paid off. We eventually got to the point of being side by side, and she would allow an occasional pat. My big reward was when one day Baghdad, completely to my surprise, put her head in my lap. That was my first big thrill of feeling close to her. It was the start of Bag love.

Quiet times with Baghdad are some of the most special moments I've shared with any animal. Establishing that special time took patience and consistency. I never interfered or interrupted her quiet time, but slowly made her comfortable enough until many times she preferred being with me during these restful times.

Even if your cat is not a natural lap cat, you will still be able to establish and share special quiet times together. No matter what your cat's temperament, you will be able to "make a cozy cat" by being patient and using the same method I used with Baghdad.

"LEAVE ME ALONE" SIGNALS

Baghdad slowly walks over to the shady area where I am standing. She looks up at me with dozing eyes, then looks down to the ground. She circles me once, then looks up and down again. I've been through this routine so many times with her, I know exactly what she's up to and gladly sit down to share some quiet times with her. She always has to position herself just so. After much preparation, she finally settles down with her head resting on my lap.

This is a nice moment of quiet time bonding, so I stroke her and speak softly to her. Life is great. She came over to me for attention, all this wonderful interaction is happening, right? *Wrong!* The next moment she gives me the kind of look I used to get from my mother that says, "If you say one more word . . . " Baghdad is telling me: not one more softly spoken word, not one more stroke on the back. At this moment she simply wants a nice soft, warm pillow. Never argue with a tiger on your lap.

You need to recognize "leave me alone" signals from your own felines. These warnings can be very subtle, or like Baghdad's, very obvious. The more time you spend with your cat, the easier it will be to read her moods and know when she wants you to interact with her during quiet time, and when she doesn't. Some signals your cat may use to let you know she wants to be left alone are twitching her tail, flattening her ears, or moving to a different spot.

FELINE PHYSICAL DURING QUIET TIMES

Quiet time is also a perfect opportunity to give your cat a checkup. He's calm and used to your touch during this time. There are many health problems that cannot be detected simply by observation. Your cat won't always let you know if he is injured or sick, though there are many signals to read his health you can pick up by touching him. Your cat can hide discomfort and pain; a gentle physical allows you to find early signs of a problem so you can alert your veterinarian. Since you spend the most time with your cat and know what is normal behavior for him, you are your vet's most reliable source for your cat's health. Follow this "feline physical":

First, *observe:*

- **Eyes** should be alert, quick to react, and have no discharge.
- **Ears** should be clean and show no signs of infection or flea infestation.
- **Fur,** the overall coat, should look healthy.
- **Mouth** look inside and check teeth and gums. The gums should appear a healthy pink color.
- **Nose** should be clean and have no discharge.

- **Paws** should not have cuts, rawness, cracking, or dryness on the pads.
- **Claws** should not appear outside the toes when retracted. If they do, they are too long and need to be trimmed.

> If you have an indoor/outdoor cat, pay special attention to bite (puncture) marks, which she may have received during a cat fight. These wounds may need medical attention to prevent infection and abscesses.

Second, *feel:*

- **Fur** should be soft to the touch. Starting with the head down to the tail, feel for sores, dry patches, scratches, cuts, lumps, hair loss, infestation, and scabs and dirt left from fleas.
- **Abdominal area** should not have any lumps. Check closely.

Third, *listen:*

- **Breath** should not be labored and should sound clear.

Your cat will probably enjoy this time of attention and will never guess your intent. For him it won't be so different from Mom's bath and grooming. With the tigers, the "feline physical" became a daily routine. If we noticed any sensitive areas or scratches, we would take note of them and watch them for a couple of days.

By establishing quiet time when your cat is comfortable with your touch, even in ill health he will allow you to be close so you can offer any treatment or assistance. By giving your cat a "checkup" during your daily quiet time, you will be able to maintain a high-quality health care program and be able to detect any problems early on. Remember to take notes and notify your vet.

QUIET TIME AND GROOMING

Sharing quiet time is also the perfect time to groom your cat. Grooming cuts down on shedding around the house and enhances your cat's health. If you have a long-haired cat, it is a must to brush her daily to get rid of knots, tangles, and any foreign objects that may have been picked up outside.

Once you've established a quiet time routine with your cat, it's easy to incorporate grooming into this time. Your cat will enjoy being brushed but you want to start out slowly. Brushes or combs scare some cats at first. To familiarize your cat with a brush or comb:

- Start slowly stroking your cat with your hand in the same pattern you would with a brush. Your cat should enjoy this and if she has any sensitive areas she will let you know with a harsh look or by flinching and resisting being touched in a certain area.

- Next, bring out a brush (a smaller brush is less intimidating) and allow your cat to investigate it while you're stroking her with your other hand. She will soon associate the soothing feeling of your stroking hand with the brush.

IN THE WILD, LONG-HAIRED CATS SHED ONLY IN THE SUMMER. DOMESTIC LONG-HAIRED CATS, HOWEVER, SHED YEAR-ROUND BECAUSE OF THEIR ARTIFICIALLY LIT AND HEATED ENVIRONMENTS.

•

- Acquaint her with being brushed by starting with light pressure. Once your cat is comfortable with the use of a brush, you may apply more pressure. Try to cover the entire body, leaving the sensitive areas, like the underbelly and feet, for last (check your cat's sensitive areas visually first for any cuts, bumps, or scratches).

- At the beginning, short brushing periods are best, ensuring her enjoyment of this new activity. You want to stop while she still appreciates being groomed—she will look forward to it so much more the next time. Look for signs of restlessness and quit before you no longer have the option.

- Once these steps are achieved, work grooming sessions up to fifteen- to thirty-minute periods. If you have a long-haired cat, these grooming sessions should be done twice a day. If your cat resists her grooming ritual, use a smaller brush and slower motions. Let her see the brush while stroking her until she is comfortable. Remember that *patience* and *consistency* are the keys.

QUIET TIME AND FLEAS

Quiet time is also the best time to deal with flea problems. This task can be as enjoyable to your cat as grooming is. The best way to get rid of fleas is by using a flea comb. Harsh and toxic chemicals never really get rid of fleas, but simply move them around. The best flea comb that we've found is called Fleamaster, distributed by Breeders Equipment Co., and most pet stores carry it. The advantages of a flea comb are many:

- It's a onetime investment that will last you for years.
- You and your cat can enjoy ridding fleas just as you enjoy grooming.

- There is no need for repeated baths and shampoos, which are usually not welcomed by your cat.
- There is no need for toxic sprays, powers, dips, and flea collars.
- There is no danger of fleas developing immunity to it.
- There is no need to vacate the household.
- There is no need to spray the yard with chemicals that harm the environment.
- There is no need to worry about flea eggs, larvae, and pupae. You simply catch the fleas and kill them every day and eventually there won't be any eggs, larvae, or pupae.
- There is no need for repellents, which only chase fleas from one place to another (and you never really get rid of them).
- There is no danger from any allergic reactions.

How do you use a flea comb? Once your cat has the grooming ritual down, follow these simple steps:

- Comb your cat from front to back starting at the head.
- Press firmly on the comb to make contact directly with the skin surface where the fleas are feeding.
- The fleas will be caught between the teeth of the comb where they can be easily pushed off into a pan of water.
- Use the flea comb daily.

By using a flea comb daily, it is possible to control completely any household flea infestation—as soon as they hatch, all fleas seek out their natural host, the dog or the cat. Your household pet is the only place you can be sure of catching them at any time. The next best method to get rid of fleas thoroughly without toxic chemicals is to vacuum your entire house daily and immediately get rid of the vacuum bag, as fleas can crawl back out.

THE BENEFITS AND STRESS RELEASES WE GAIN FROM SHARING QUIET TIME

With today's often busy and stressful schedules, sharing quiet times with our pets gives *us* many health benefits. We all know now the health hazards of overly stressful lives. When you share quiet time with your cat, you are also giving yourself time to unwind and relax. There have been many studies and successful programs using "animal therapy" with elderly and differently abled persons. These programs can be extremely effective. We should also be aware of the therapeutic effect our pets have on all of us. Whether by sharing a quiet time by the fire, an active game of chase, or just providing the chuckle we get watching some comical move they make, our pets enhance our everyday lives. Don't forget to thank your companion every now and then for doing so much for your own well-being.

MEDITATION WITH—AND FOR—YOUR CAT

You will find that just taking as much time as you can on a daily basis to interact with your cat will relieve some of the tension and stress of the day. The first step to relaxation is to turn away from external distractions. Sit alone quietly with your cat, with no distracting stimuli from the outside world. Sit in a comfortable position with your posture straight. Focus your mind on a single thing, like the sound of your cat's purring or breathing, for concentration. Focus on the present moment to help yourself concentrate; this leads to a state of meditation. Count the number of purrs your cat purrs in one minute. Now count your own breaths. How many seconds does it take for you to inhale and how many to exhale? This time for observation readies you for a deeper state of meditation.

CATS AND OTHER PETS HAVE BEEN USED SUCCESSFULLY TO ACHIEVE POSITIVE REACTIONS FROM AUTISTIC CHILDREN.

•

Now sit quietly listening to and watching his breathing for another five minutes. Once again count how many seconds it takes you to inhale and how many to exhale. You may be surprised to see that simply by concentrating on your cat's breathing, you've lengthened your own breath. You may not have nine lives, but the one you have will be longer. Yogis believe a long breath means a long life—and a healthier life. Those who practice meditation regularly achieve a slower heart rate, a more stable respiratory rate, lower levels of anxiety and fear, and even lasting, more loving relationships.

NUMEROUS STUDIES HAVE BEEN DONE PROVING THAT TOUCHING AN ANIMAL *LOWERS* BLOOD PRESSURE, WHEREAS TALKING TO AND EVEN TOUCHING ANOTHER HUMAN BEING *RAISES* BLOOD PRESSURE.

●

PARTNER BREATHING WITH YOUR CAT

Some schools of psychotherapy believe that disharmony between two people arises because one person's breath is out of synchronization with the other person's. Partner breathing—in which one person breathes and the other "follows" the partner's breath—helps dissolve the barriers between people and brings them closer to each other. You can use this same calming technique with your cat during quiet time.

First sit and quietly listen to and watch your cat breathe. This will help you to train your attention and learn to concentrate and listen—all necessary steps for meditation. If your mind feels restless or your attention wanders, simply bring your attention back to your cat. See if you can watch and listen to her breath for a full minute. Then see if you can continue for two while you synchronize your breathing with your cat's.

Any number of studies have proven that deep breathing effects a relaxation response. In fact, one of the most important parts of

Hatha Yoga is *prānāyāma*—breathing practices aimed at controlling breathing patterns in order to soothe the nervous system and balance the emotions. Here's something you can do instantly to set up your own relaxation response:

- **Position**   Lie on your back with your legs up on a chair seat. Put a pad under your head. Place your cat on your abdomen. (Caution:

Do not do this if you have had recent abdominal surgery, are pregnant, or have had gastrointestinal problems . . . or if you have too large a cat!) Relax your face, neck, and shoulders.

- **Breathing** Inhale into your belly—feel your back broaden against the floor. Gently, exhale slowly. (The exhaled breath is the important breath here. It is often called the "relaxing breath," while the inhaled breath is often called the "energizing breath.") Now place the cat on your chest and inhale into your back. Continue to inhale as you watch your cat rise with the inhalation. Slowly, gently exhale. Watch your cat fall with the chest. Now take your cat off and continue inhaling and exhaling. The weight of the cat will have made you more conscious of the rise and fall of your abdomen and the widening of your back and chest. Now inhale into your back and abdomen and slowly exhale. Breathe normally and observe the fullness of your breath.

- **Visualization** Picture yourself as a big cat lying on the floor, belly and spine relaxed. Now watch yourself as that imaginary cat breathing so generously you can see your body expanding with each inhalation and contracting with each exhalation.

    The feet up on the chair relaxes tired legs and helps the blood in the legs return to the heart. Being conscious of your breathing allows you to reeducate your breathing pattern. This relaxation exercise is wonderful before going to bed, or when you are tense.

- **Relaxation** During quiet time with your cat, put some soothing music on and pick up a book, sketch pad, needlepoint, or any other project that is relaxing for you to do. Put your feet up and sit next to your cat or with your cat on your lap. You will find the soft purring sounds your cat makes very calming.

THE NEXT TIME YOU ARE ABOUT TO HAVE A BIG DINNER, SPEND 5 MINUTES "GRACE TIME" BEING QUIET WITH YOUR CAT *BEFORE* YOU EAT. IN A RELAXED STATE WE USUALLY EAT LESS, AND MAY DIGEST BETTER WHAT WE DO EAT.

•

These methods are just a few of the ways you can relieve daily stresses while spending time with your cat. You will also be able to explore your own ways to relax with your cat by simply spending quality time together.

you and your indoor cat

. . . a morning kiss, a discreet touch of his nose
landing somewhere on the middle of my face.
Because his long white whiskers tickled,
I began everyday laughing.
Janet F. Faure

THE GREAT INDOORS

The head moves slowly back and forth, eyes wide, looking for the perfect spot. The moment is right and Penguin launches herself on to the bookcase. Papers fly and books fall over, but that doesn't matter, because for some reason known only to her, she had to be precisely there at that moment.

Some cats are indoor cats, some cats are outdoor cats, and some cats are indoor/outdoor cats. If you live in a big city or anywhere the outside elements are too dangerous for your cat to go outside, then it is safest for him to spend the majority of his time inside your house. If your cat spends all his time indoors or half of his time inside and outside, you need to learn some methods so you can coexist harmoniously under one roof.

Cats get themselves into trouble indoors because of the objects (vases, lamps) that get in their way when they are showing natural behavior. Other times it's that pent-up energy that gets them into house-damaging predicaments. Still other times it's "cat curiosity" that results in a broken dish.

All cats, young and old, seem to have impish ways. By learning a few tricks, channeling your cat's energy, and applying the basics of conditioning, you will be able to teach your cat the behavior necessary for a happy indoor life together.

CAT BOX AND SPRAYING

A tiger's litter box in the wild is as big as his territory. Of course your indoor cat's litter box is much, much smaller. Cat owners bring up the toilet habits of their cats to us only when they're having a problem. We've seen cat owners become very tense over their pet's hygiene problems, causing their cat to become stressed too. If your cat is having difficulties, like not using her litter box, take a look at possible reasons, try the following procedures, and always remember the basics of conditioning.

Indoor cats, outdoor cats, all cats need a litter box. Litter box training for your new kitten or cat is simply starting a new routine. You can litter-box-train a cat at any age and you should start immediately after you get your new cat home. You should also be sure to use environmentally safe cat litter (see Chapter 10).

Placing the Litter Box: At the start, place the litter box in the room where you will be training your cat to use it and where he spends most of his time. Once he is consistent, you will be able to move it to an out-of-the-way corner—in the laundry room, mudroom, bathroom, back porch, basement, etc.

Small Area: While training your cat to use a litter box, keep her in one room until she is comfortable using it. If she has full rein of the house, it will be more difficult to get her to focus on the job she must do and the small area to which she must always go.

Start Simple: Use plain litter in the beginning stages, since deodorized litter may not be to his liking or may confuse him. Purchase a low-sided box without a cover for easy access. Don't add anything that might frighten him.

After a Meal: The best time to introduce your cat to her new litter box is after a meal. Gently place her in the sand. Usually nature will take over. If not, be patient and give your kitten or cat some time. You can gently move her paws in a scraping motion to motivate her.

Oops, Wrong Corner: If your cat or kitten makes a mistake and goes outside of the litter box, do not yell at him (remember the basics of conditioning). Do not rub his nose in the "mistake." This will only confuse and excite him, which is not good for training. Move the litter box to the spot and, if possible, put the evidence of his mistake in the box. Then gently place him in the box. Do this consistently every time he goes outside his litter box and soon he will get the idea he must go inside.

Cleaning: The number one reason a cat does not use her litter box is because it is not clean. A cat will stop using a dirty litter box. Cats spray only to mark territory; the rest of the time it's instinctual for all cats to hide or mask their scent, and that means covering it up. If the litter box is not kept clean—if there is too much waste or the smell of urine—your cat's natural instinct will be to find another place in the house. The litter box cleanup routine is not an unpleasant chore if performed on a daily basis. Here are some hints:

- You need a litter scoop, a pail with a cover, and disinfectant.
- Store tools close by so it will be more convenient for you to get into a routine.
- Remove any waste matter twice daily and put the waste into a sealed container. If you use environmentally safe litter products, it's even more convenient because you can flush the waste down the toilet.
- Once a week scrub the litter box with disinfectant and replace old litter with new. Remember to rinse the disinfectant out thoroughly, or the smell could deter your cat from using the litter box.

Litter Box Privacy and Location: The final destination of your cat's litter box is important since cats, being creatures of habit, will not follow the litter box around. Once you've established the use of the litter box, placing it in a permanent spot is essential. For your cat to feel comfortable and have some privacy, find a spot in your house that is easily accessible so that your cat can get to it at all times; that is a low-activity area without noise, kids playing, or anything unsettling that would cause your cat to go elsewhere; and that is your cat's individual spot. If you have more than one cat, don't assume they want to share. Many times one cat will intimidate the other. If this is the case, get another litter box and place it in its own spot, some distance from the other one.

LITTER BOX SHARING

If you have more than one cat, you will want to start each off with her own litter box. Each should have her own spot apart until both are very comfortable using a litter box. After that point here's how to teach them to share one:

- Slowly move the litter boxes nearer to each other, giving the cats time to feel comfortable with their litter boxes closer together.
- Then move the litter boxes side by side.
- Help them get used to a new scent in their box by switching a little litter from one box to the other.
- Keep the boxes in this position for a while until you notice your cats using *both* boxes.
- Be watchful for changes in behavior such as refusal to use a box, finding a new spot, or aggression. If behavior problems occur, go back to the previous step where there weren't these problems and go at a slower rate.

Remember, it's much more important for them to have good litter box behavior than for you to have the convenience of one litter box! Even when your cats are comfortable sharing, it's always good to have an extra litter box on hand in case one of your cats get sick and needs to be separated from the other.

YOUR INDOOR CAT AND SPRAYING

Spraying is much different from indiscriminate urinating. It has a definite purpose and is a very difficult behavior to break. Nonneutered males spray as a means of marking territory. Sometimes even neutered males and females will get into bad spraying habits because of stress, boredom, or something or someone that seems to threaten their territory. Spraying to mark territory can also be triggered by a cat that your pet sees outside through a window.

Since spraying is a behavior, your cat can be conditioned to stop. Try these preventive steps:

- **Neutering**   If your male is not neutered, call your vet immediately and make arrangements to have this done. This will also help avoid unwanted cats in the future.
- **Increased Activity**   Spend more time with your cat, increasing playtime and activity, with the purpose of dispelling her boredom.
- **Stress Signals**   Look for indications and behavioral changes from your cat signifying stress. For example, is he spending more time by the window? Is he startled easily?
- **Stress Release**   By identifying the cause of your cat's stress, you will be able to relieve it. Spend more time with your cat and try to eliminate any outside stimuli that could be causing the stress.
- **Stopping Threatening Actions**   Anything that could be perceived as a threat to your cat's territory will cause him to spray. If another household cat is a threat to your cat, then keep them in different rooms.

These steps, applied with the basics of conditioning, will stop your cat from spraying. But what do you do in the beginning when you find a "sprayed spot"?

- Clean the sprayed area immediately with a strong disinfectant. If you don't, your cat will continue to spray and develop a stronger habit.
- Make the spot unappealing to your cat. Put aluminum foil on the spot and lay wax paper on the ground by the spot—these make

YOU CAN TELL THAT A CAT IS SPRAYING IF HE DOES NOT SQUAT, BUT STANDS WITH HIS TAIL HIGH AND ACTUALLY AIMS FOR HIS TARGET.

•

unpleasant sounds when he sprays or walks in this area. Also, two-sided sticky tape laid on the ground in the area sprayed will be unpleasant for him to walk on.

- Place a food dish by the sprayed area. This will discourage your cat from spraying there.
- A firm "no" should be said when he is caught in the act. Never rub his nose in it or yell—the objective is to teach, not to punish or startle him.

We don't have to condition tigers against spraying because it is important for their social structure, and their environment can handle it. The big cats enjoy the strong odor—much more than the trainers do.

HOUSEHOLD HAZARDS

When you have a pet cat in the house you need to be extremely careful of household hazards that could cause her damage. Here is a checklist of safeguards to follow as you survey your house:

CHECKLIST

❑ 1. Cats are attracted to **SHINY OBJECTS**. Be sure that pins, needles, earrings, and all small items your cat could swallow are out of reach.

❑ 2. **ALUMINUM FOIL BALLS** are a fun toy for your cat but should be used only in a tight ball and under close supervision. If swallowed, a foil ball can cause intestinal blockage.

❑ 3. Never leave **PLASTIC WRAP** or **PLASTIC BAGS** around, since your cat could accidentally suffocate while playing in it.

❑ 4. **MOTHBALLS** should always be kept in *sealed* containers and out of your cat's reach. All cats love to play in dark hidden places, the same places you may store mothballs. Mothball fumes will cause liver damage in your cat if inhaled.

❑ 5. **CARPET CLEANERS** and **FURNITURE POLISHES** can leave a low-lying toxin in the air that can be harmful to your cat.

❑ 6. **CLOTHES DRYERS** are warm and inviting to your cat. Always check inside *before* closing the door.

❑ 7. The **TOILET LID** should always be kept shut to prevent a panicked kitten from drowning.

❑ 8. **WINDOWS** should be kept secure if you live a few stories up and your kitten or cat could be in danger of falling.

❑ 9. **ELECTRICAL CORDS** are very dangerous for an unconditioned cat who likes to chew. Until you condition your cat to stop chewing, make sure your electrical cords are not hanging and secure them to baseboards. Tabasco or lemon rubbed on the cord will also keep your feline from chewing it.

❑ 10. Make sure **TRASH** is well out of reach so your cat does not swallow anything harmful.

❑ 11. **SMALL APPLIANCES** such as irons, toasters, and curling irons can burn your cat if they accidentally connect. Make sure you turn these off and they are out of reach.

This is a general checklist, but remember, cats are always finding themselves in peculiar places and circumstances. Always keep your cat's safety in mind and never shut a refrigerator door or any drawers without first checking for a curious cat.

THE COUCH ISN'T A SCRATCHING POST

Scratching is important for maintaining claws. When a cat scratches, he's not honing the sides of his claws, but rather pulling off the outer layers. When the older portion sloughs off, it reveals a new healthy claw. Tigers can slough off a two-inch sheath, leaving them with a smaller and sharper claw. But why does a cat go straight to the couch or your favorite chair and start to scratch in the most contented manner? The two are like magnets—no matter how much you fuss and say no, the scratching continues. And why do cats always prefer the best furniture when they have a brand-new scratching post just sitting there unused? This is one of the most common household pet problems and a question we are asked regularly. With the high cost of furniture, this concern is entirely justified!

Before Bill finally figured out the solution for this problem he was very frustrated. In despair he looked at Rakon, hoping for an answer. Rakon yawned. With a tiger, it's not just a matter of scratching the couch, it's more like grabbing hold with his claws and running off with it! On this particular day, he went home from work with the scratching dilemma on his mind. Using the basics of conditioning, he started to *think like a cat* and got down to floor level to look up at the couch. (Judy, quite used to this behavior, just handed Bill a dust cloth and asked him to get the corners while he was down there.) From this vantage point, from a cat's perspective, he could see how a couch could look inviting. It's the perfect height for a cat to stretch and extend himself, and gives him a better angle to pull his claws through the fabric.

Okay, he could understand why a cat needs to scratch to keep his claws sharp; it feels good. But why the couch instead of the scratching post? Then it dawned on Bill: The first thing a tiger does when approaching his territory is to mark it by spraying and scratching. Scratching marks territory by leaving his individual scent from the sweat glands of the paw pads. Bill then realized that all domestic cats do the same thing, and it doesn't matter what the furniture is—it's where it is that's important. The first thing your cat does when entering his territory is to go over to a piece of furniture and leave his signature with his claws. This is usually near an entrance—front door, back door, sliding glass door, etc. The *nearest* piece of furniture is the target for your cat to scratch in his signature. *Eureka!* If you want your cat to use his scratching post, you need to move it out of the corner, out of the back room, and into his "first approach" territory. This is all you need to do to stop your cat from scratching your furniture:

- Determine the traffic area where your cat typically enters his territory.

- Make or purchase a scratching post. The most desirable store-bought scratching post for your cat is covered with sisal. You can make your own by attaching a four-by-four piece of wood onto a sturdy base and covering it with the *back side* of carpeting. Make sure the post is taller than your cat's full length so he can get a good stretch. Be careful that the post is strong and does not wobble or bend. Make sure there are no sharp points or nails sticking out. The material covering the post should be strong, but not such a thick mass that claws can get stuck.

- Place the scratching post near the entrance of your cat's territory. If he has a long record of scratching the furniture, then put the scratching post right near the scarred piece. You can rub onto the post either small amounts of catnip or your clothing with your scent, making it more desirable than the furniture.

- Always praise your cat after he uses the scratching post, particularly in the early stages.

- After your cat consistently uses the scratching post, you can move it to the most convenient place in the room. Do not try to move the post too soon. Slowly move it six inches at a time toward the desired area, leaving it at its new spot a few days at a time. Once your cat is comfortable with his own "marking post," he will go to it even if it is not near the entrance of his territory.

SINCE AN INDOOR CAT'S SCENT WILL DISSIPATE AND DIFFERENT SMELLS WILL COME INTO THIS AREA, HE WILL CONSTANTLY NEED TO REESTABLISH HIS INDOOR TERRITORY.

•

THAT WAS MY FAVORITE VASE!

As writer Wesley Bates says, "There's no need for a piece of sculpture in a home that has a cat." Cats are agile creatures and are usually careful when they walk, but not even a favorite vase matters when they're on the trail of a fly. There are many ways *you* can avoid damage to your favorite household items. Here are several:

- If your cat's favorite sunning spot is near a window where you display knickknacks, you will be happier finding a different location to showcase them.
- Take note of your cat's favorite investigating and hiding spots (remember the predator likes to hide when stalking her prey). If you have favorite breakable items in these areas, try finding a new place for them.
- All breakable items should be stored in areas that can not be reached or easily knocked over. Playtime sessions will rid excess energy that can cause damage.
- Remember to think like a cat and look at the problem from her angle. Sometimes the easiest solution is removing the problem.

PLANT VANDALS

Some cats have a tendency to vandalize certain types of indoor plants. They may stomp, dig, chew, or scratch your prized vegetation. Your cat usually doesn't mean any harm to your plants; the fact is, he's probably very fond of them. The aroma, the feel, and his instinctive allure to vegetable extracts make your plants very tempting.

You can use the basics of conditioning to control your cat from damaging your plants, but you can also grow your cat his own little

garden! This is great if you live in an apartment or big city and your cat doesn't have regular access to the outside. Get a small seed box and plant it with some of these herbs and weeds: catnip, thyme, sage, parsley, lawn grass, chickweed, coltsfoot grass, and other types of grass like wheat or oat. Your cat will appreciate your efforts to supply him with his own plants to do with as he chooses.

This special garden will also divert him from *your* house plants. Some house plants are poisonous to your cat, and you need to be very careful that your cat does not have access to those plants, the most common of which are caladium, dumb cane, poinsettia, ivy, mistletoe, oleander, common or cherry laurel, azalea, and winter cherry. Make sure all poisonous plants are well out of your cat's reach or better yet out of the house.

Sometimes the simple solution is to move the plants that might be in your cat's path or in a favorite resting or sunning spot. If they're off the path, they may not be as fun to play with. If you don't move them, rubbing Tabasco sauce or lemon juice on their leaves is a great technique to stop your cat from chewing on them. And remember, active time will relieve your cat's boredom and cut down or cut out plant damage.

Can you stop all cats from biting plants? Well, some can be stubborn and in the end your cat may learn, but you may not have any plants left.

Mohan was a good example of that. When Mohan was moved to Vallejo, he took a special interest in some large, mature willow trees and started taking huge chomps out of them. We began to condition him against this behavior, but he was persistent and would sneak a bite every chance he could get. The trees couldn't take much more of Mohan's "appreciation," so it was time to take action. We covered the trees with wooden corsets; these not only look nice, but the tigers could also use them as scratching posts. Moral of the story? In stubborn cases, look for alternative solutions—be creative.

FEEDING FACTS

Your indoor cat or your indoor/outdoor cat will have certain foods she wants and she will give you "how it's to be served" rules to follow. Here are the basics:

- Fresh food is definitely preferred.
- Food must *always* be served from a clean bowl.
- The clean bowl should be placed in a quiet, calm area that isn't too bright.

We often get asked how much food to feed a cat and how often. Well, your cat may tell you *a lot* and *often*. But many times listening to your cat on this subject results in her obesity. Because each cat is different in body structure and personality, devising a feeding schedule is not always simple. For best results, watch your cat's behavior. Is she very active and eager every time you enter the kitchen? Or is she sluggish at feed time? Her mealtime activity level can help you judge how much to feed her; the more active she seems whenever you get near the food, the more food she may need.

Your cat's appearance also can help you determine how much food to give her. But if there is sudden weight loss, don't just give her more food: Advise your vet right away. If your cat's activity level slows down and she starts to look "pudgy," then you know it is time for smaller portions. When in doubt, ask your vet what he thinks the normal weight for your cat should be.

FEEDING ADVICE

- Do not leave food out all day: If the food is out for a half hour or forty-five minutes, that should give your cat plenty of time to consume it. Remove any food that isn't eaten in this time period. This will ensure that your cat is only eating fresh food—and will help avoid finicky behavior.

  Removing food that is left uneaten after about forty-five minutes will also help ensure that your house stays free from ants, roaches, and other pests that like to congregate around half-eaten bowls of cat food.

- Smaller portions and several meals per day is ideal. The following is a general feeding schedule we recommend:

Age	Meals Per Day
Weaning to 3 months	4 to 6
4 to 6 months	4 to 5
7 to 12 months	3 to 4
12 months and over	2 to 3
Pregnant females	3 to 5
Older cats	3 to 6

- The amount of food you start to give your cat can be adjusted by how much he will eat in one sitting. For instance, if you start with 1/4 to 1/2 cup of food and food is continually left, then cut back. If your cat is finished in a flash and looking for more, then it's time to increase the portions.
- Cats need some dry food for strong teeth. But don't give dry food if your cat has bladder problems.
- If your cat is not eating or is finicky, try baby food. Our finicky one loves chicken or turkey. (Be sure to strain the food before serving.)

FEEDING TRANSITIONS

- Kittens should eat kitten food for the first twelve months. During this time they will require essential nutrients for strong bones, teeth, and muscle development. After one year, a steady diet made for adults will be more suitable. Transition from kitten food to regular cat food should be gradual, over five to ten days. This will help in making your cat accept a new diet.
- Since pregnant and nursing cats will need their diets adjusted to assure proper fetal development and milk production, it's best to consult your vet and develop the best changes. Again, any change should be made gradually.
- As your cat becomes older and less active, you will need to adjust food intake so unnecessary weight gain does not result.

YOUR FINICKY CAT

If your cat is finicky or has lost her appetite, look at the following list and see if any of the most common reasons may be the cause.

- Upset from a change in routine or traveling.
- Hot or humid weather.
- Female cat in heat.
- The fresh food you're serving is slightly "off" (remember, your cat's sensitive nose will smell the beginning stages of spoiled food way before you will).
- The nearby restaurant's or your down-the-street neighbor's menu is more appealing to your cat than yours.

If none of these explanations seems to apply, contact your vet after a twenty-four-hour period of your cat's refusing food.

NOURISHMENT NO'S

Is there anything you should not feed your cat? Yes! Follow these "Nourishment No's":

- No table scraps—these can cause your cat to reject his proper diet and give you a finicky eater.
- No raw eggs—they contain chemicals that prevent absorption of biotin, which results in a nutrient deficiency for your cat.
- No boiled fish—boiling destroys all the valuable nutrients.
- No poultry bones—they are small and soft and can splinter, which can cause your cat to choke.
- No dog food. Pet food is especially designed for a cat or a dog. Cats require much more protein than dogs do.
- No dry food for cats with bladder problems or FUS (Feline Urological Syndrome). FUS is a condition in the lower urinary tract associated with a high urinary pH and high levels of magnesium. This is an increasing dietary problem.
- No quick diet changes. If your cat's food is changed too quickly or abruptly this may cause an upset stomach or loose stools. Any changes should be introduced slowly and gradually. Mix a little more of the new food with the old each day until the food is all new.

The best thing to remember is that the fresher the food you feed your cat, the more nutrients she is getting. As always, the more time you spend with your cat, the more you will understand her needs.

TIGERS EAT NOT ONLY LARGE PREY BUT ALSO FEED ON FISH, FROGS, GIANT MONITOR LIZARDS WHICH GROW AS LONG AS 9 FEET, AND PYTHONS WHICH CAN GROW TO OVER 20 FEET LONG.

•

KITCHEN AND DINNERTIME TIPS

Cats in the kitchen can be a test of patience—especially if they are constantly under your feet with a stout cry of "I want food." They can be demanding and at times so bold as to hop on the counter and take what has not been offered to them. Why is it that cats lose control at the mere thought of food? Think back to the predator. Cats are excited by food, and the more they think about it the more excited they can become. If you want a controlled cat in your kitchen, you must be consistent when dealing with food.

House cats do not get aggressive around food but they do get excited. Tigers, on the other hand, become excited and work their way to aggression. Cats don't intend to irritate us by getting underfoot (they don't want to be stepped on), but if they are not taught control, they can be troublesome in the kitchen.

We've taught the tigers to be in control when eating around us. Judy had always seen Mohan (the big male tiger) in control. He would do something simple, like standing up with his front paws on my shoulders, and when he was finished I would reward him with a piece of meat. He is always gentle when taking meat from my hand and it always looks very controlled and relaxed. But later I took Judy back to see Mohan eating his dinner. He was anything but the gentle tiger she had seen hours before. Anticipating his food, he becomes excited, taking quick strides back and forth. From a distance I throw him his ten pounds of horsemeat; immediately he is all teeth and claws, growling and roaring. This was Judy's first time seeing the tigers at dinnertime and it surprised her just how excited they get. When I feed the tigers their dinner, they are able to get excited and display their natural instincts in their own environment. When they are with us and offered a food reward they have to stay in control.

You must set a controlled routine for your cat in the kitchen, and be consistent. Following are typical kitchen and dinnertime problems and what you can do to solve them.

Owners' Regulations: To cut down on bad kitchen habits, you can start by doing your part. Think ahead and you will be able to help your cat take heed against temptation.

- Don't leave edibles out on the counter.
- Shut all pantry doors, especially where you store cat food.
- Put dry food in sealed containers (this will also make the food last longer).
- Place the trash can in a location where your cat cannot get to it.

Cat Under Your Feet: If you have an exuberant cat darting back and forth, one who just can't wait for his food and is developing bad begging habits, you need to teach him not to associate the kitchen with food. You can do this by following these three simple tips, and remember, **A**lways **B**e **C**onsistent:

- Do not give treats in the kitchen.
- Never drop food on the floor as a treat.
- Give treats or dinner in one area only, away from the kitchen.

These tips will help your cat to associate mealtime with a different area and keep him from being an irritant when you're in the kitchen. By never feeding your cat in the kitchen, only in his food dish in another room, you will be able to keep the kitchen as *your* food place. If you want to give rewards for good behavior, do it in other rooms. By consistently feeding or rewarding elsewhere, your cat will not be under your feet in the kitchen, hoping for a quick treat.

Roaming the Countertop: Jumping on countertops is a bad habit that comes from wondering what's going on up there. Your cat sees the family constantly fiddling around on countertops and wants to know what's going on. So up she jumps, right to the source, especially since she suspects the counter may have something to do with F-O-O-D. This should not be allowed. To keep your cat from jumping up on the counter:

1. Feed your cat only from her food dish so she will develop the habit of eating out of it only and not from other surfaces like countertops. (This also discourages your cat from eating something harmful somewhere else.)

2. Paws are not allowed up against the sides of the counter. This is phase two of getting to the countertop. After she has sat on the floor, staring at you, with her paws up against the counter, all it takes is a quick jump—and there she is, right where you don't want her. Take notice and if this happens, give a definite "no" to discourage her. Be consistent!

3. Never allow your cat on the kitchen table. This is an area also associated with food. Treat it as another countertop—no jumping onto it.

4. If you see your cat on the counter, do not overreact, but deal with it quickly—with a firm "no." If she does not obey by jumping off, then repeat your "no," picking her up and firmly placing her on the ground. Don't shove her off the side. You are the mother figure and must give your cat direction, not scare her off.

Continue these methods even if your cat is stubborn, and stick with the basic rules of conditioning. If she is persistent, try

making the countertops a bit unpleasant. Something that feels unstable or makes noise will cause her to avoid jumping there. Sticky tape or a newspaper over a set mouse trap (please be absolutely sure it's upside down) works well to discourage your cat from going into *any* area you do not want her in.

Cat in the Cupboards: Cats can take the initiative and look for food all by themselves. They *can* open doors, pull cupboards open, and even get kitchen drawers open. One of the biggest mistakes people make when they see their cat in the act of trying to get into a food area is to watch and see if he can do it. They are surprised that most times he can figure it out and a habit is formed. Instead of just watching, you need to assume your motherly role and immediately give a firm "no" to let him know this is not acceptable behavior.

While you're in the process of conditioning your cat to stay out of cupboards and not open cabinet doors, consider using baby-proof latches. These are hard to figure out and will give you a chance to correct your cat's behavior before bad habits are formed.

CAT CALLS

Writer Arthur Weigall says it best: "If I called her, she would pretend not to hear, but would come a few moments later when it could appear that she thought of doing so first." People ask Bill if tigers come when they're called. Yes they can, but it has to be for a good reason like a greeting or a food reward. A dog will usually always come when called, whereas a cat most times wants to know the reason first. If you share quality time with your cat and have established an enjoyable playtime and quiet time with her, then she will be much more responsive when being called.

All cats *can* learn to come to a certain call or signal.

All the tigers respond to set signals we use to call them. This required a lot of training and the amount of time it took to train them differed depending on the individual tiger. Rakon, for instance, learned quickly and always lifts his head and turns

toward us when we call his name—he knows it's usually a warning that his sister, Kali, is sneaking up on him.

We have conditioned the majority of the tigers to come when they see a clenched fist. If they are far away we raise the fist up over our heads. If they are near, the fist goes right in front of our body so they can see it. We trained them to respond by using food rewards some of the time, but using praise always.

When you are first teaching your cat to come to a certain call:

- Establish a signal to use along with saying your cat's name: a whistle, snap of your fingers, clap, anything that will be easy for you to use consistently and for her to come to continually. The problem with using only your cat's name when calling her is that it is a word she hears constantly and it soon loses its effectiveness and can confuse her. Adding a signal when you say her name should eliminate any confusion about what it is you want. With the tigers, many times we will use the signal only to call them. You may want to try doing this with your cat.

- Start right next to your cat, say her name and give the signal, and wait for her to respond. Once she consistently reacts to the signal, slowly increase your distance.

- At first, give her a food reward *and praise* every time she responds. After it becomes a habit, use fewer food rewards but still lots of praise. (Remember, when your cat does what you have asked of her, reinforce her with a tasty treat, affection, play, or a favorite toy.)

Upward Ho!:   Once you've established a "cat call" to which your cat responds, you can teach him to jump onto a chair or a set spot. This is useful to teach him to stay in his spot while his supper is being prepared, to keep him from being underfoot in the kitchen, to prepare him to get into his traveling box, to stop begging behavior, or simply to hop up next to you for a friendly pat. Here's how:

- Give the signal for him to come over.
- Start with a chair and tap the seat where you want him to jump. He will come over to investigate (maybe thinking you have food).
- Reward him when he jumps up onto the chair.
- Call him down and start over.
- When he consistently responds to the tapping, give the exercise a name like "up," and pair the tapping with the word.
- Eventually discontinue the tapping and your cat will respond to the word command.
- Take your time and be sure to stop before he loses interest. Give him a lot of praise during and at the end of the session.

Cat Greetings at the Door: There's nothing like a welcome greeting from your cat after a long day of work. This is simple to teach because once you've established a secure relationship, she will be happy to see you and seek out your company. If she tends to be a little more distant, she may need a little more incentive. To do this:

- Give your cat a treat to get her attention.
- Rattle your keys and give her a reward.
- Then go out your door and again rattle your keys. Come back inside and immediately give her a treat.
- Repeat this several times.

- Keep rewards with you at first and give your cat a quick reward every time you enter the door.
- Remember to take a time-out if your cat is losing interest or becoming frustrated.
- Then praise your cat and offer food rewards. At first give treats every time you come in the door, then offer them randomly. The end result will be a warm welcome home at your door.

This also works well with outdoor cats to call them home. You can use any signal you like. I've been told of one cat who came to the honk of the car horn. Although this worked well for that family, I suggested changing the signal, because the association with cars could be dangerous and the cat was walking to the car instead of the house. The signal was changed to a whistle in front of the door. This transition was easy and happened quickly because the cat couldn't wait to see her mother figure.

CAT TREATS

Most cats are very food-motivated and you can use this to your advantage when conditioning your cat to do something. There are many treats on the market, so it is a matter of choice and preference. What kind of a treat reward to look for:

- Small and bite-size, not something that takes time to chew, as this will interrupt the cat's concentration. It is very important in the early stages to keep a cat's attention.
- Easy to handle, not something oily and that sticks together. If you can't hold on to the reward and are dropping it, it will slow you down and confuse your cat.

- Nutritious, something that's good for your cat and a part of a well-balanced diet.
- Dependable, something that lasts and keeps well. You want to know that the reward is fresh and hasn't become rock-hard or spoiled.

You will want a variety of treats to keep your cat's interest so he doesn't take the attitude of "Oh GRRRRReat, another tuna puff."

First find out if your cat likes the treat by checking his response to the treat *before* you ask him to do something. If he does, stop and save it for reward time. Experiment with different treats until you find your cat's favorites.

cat's night out— your independent indoor/outdoor cat

*It is in the nature of cats to do a certain amount of
unescorted roaming.*
Adlai Stevenson

THE DECISION OF AN INDOOR VERSUS INDOOR/OUTDOOR CAT

It is wonderful to have a cat share your life and home. Too, it is wonderful to let your cat go outside for fresh air and exercise. Many cats are completely outdoor cats, such as the farm cat, who is quite content living outdoors and controlling the rodent population. If your cat is an outdoor cat or indoor/outdoor cat, this chapter will help you safely introduce your cat to his outdoor environment and choose safer alternative methods to get both of you fresh air and exercise.

With the decision to allow your cat outside comes many responsibilities and concerns. A cat's nature is perfect for the ever-changing environment of the outdoors. Cats can have endless afternoons watching birds in action and chasing butterflies. They can move with the changing sun and always find those perfect toasty napping spots. They get regular exercise and fresh air. But an outdoor environment can also be harmful and even deadly to a cat.

Before deciding to liberate your cat, take a careful look around your neighborhood—soon to be his new territory. Is your house in or near a heavily congested traffic area where your cat could be in danger of getting hit by a car or truck? Are there dogs or unneutered male cats that could pick fights with him and cause injury to him? Are there many stray cats that do not look healthy and could transmit contagious diseases to him? Are there any areas around containing chemicals or poisons or is there antifreeze lying around? Is it a high-risk area for internal or external parasite infestation? The time to make the decision whether to liberate your cat to the outside world or keep him as an indoor cat is *after* you have taken this careful and complete look around your neighborhood. In some circumstances this is a very difficult decision and one that should never be made without careful consideration.

Today more and more people are making the decision to have indoor cats. The number one reason for this has been the dangerous and fatal accidents, especially those involving moving vehicles, that have occurred to cats. After assessing your own situation, if you feel that the environment outside your house is not conducive to your cat's safety, then make sure your cat gets sufficient exercise and stimulation indoors. If you feel that the environment outdoors is safe

STATISTICS FROM THE DUMB FRIENDS LEAGUE OF DENVER SHOW THAT CATS WHO STAY INDOORS ALL THE TIME LIVE *THREE TIMES* LONGER THAN ONES WHO DO NOT.

•

for your cat, GREAT! There's nothing that duplicates the stimulation a cat gets from the ever-changing environment outside.

HOW TO INTRODUCE A CAT TO THE GREAT OUTDOORS

Before your kitten or new cat puts one paw outside, make sure that all vaccinations are up-to-date and your pet has proper identification. Your cat also *must be* spayed or neutered.

Take several pictures of your cat and put them with all her papers, then store them in a safe place. This is a safety precaution in the unfortunate event that your cat is lost or stolen.

Then identify your cat with a collar and tag or a tattoo. A breakaway collar is the safest kind of collar because it prevents the cat from choking or accidentally hanging herself. However, when the collar breaks away, the identification tag is lost. For this reason, many people are tattooing identification on the inside of their cat's ears. This pain-free procedure guarantees permanent identification. For information on tattoo identification, call Tattoo-a-Pet at (1)(800) TATTOOS (828-8667). Be sure the identification includes your name, your cat's name, your phone number, and your address. The Animal Welfare Institute is doing a national study of pet thefts. You can get survey forms by writing to P.O. Box 3650, Washington, DC 20007.

When's the Right Time?: Before your cat or kitten tries a new environment, make sure he is very comfortable with his indoor home and it is a secure place for him. All the "newness" of the house should be finished: Your cat should know his way around, and be playful and investigative. His eating habits should be steady. Before a kitten is let outside, make sure you're seeing appropriate weight gain. Litter box training should be completed. Don't be in a hurry to get your cat into the great outdoors. It is vitally important that there is a secure environment in the house first. A cat's home is what he comes back to, and Mom is whom he comes back to, so the indoor home must be a positive place where he is warm, comfortable, well-fed, and secure.

The First Day Out: Whenever we first let one of our cats go outside, we feel like parents on a child's first day of kindergarten. The following tips will help relieve anxieties for both of you.

- Plan ahead of time before going outside. Pick a quiet area on a nice day.
- Stay with your cat and supervise the first few outings.
- Stay away from large things that your kitten can go under. She'll show a strong tendency to run for cover if startled or scared. Coaxing her out from under the house is not the best way to start.
- Keep some treats with you for a sure way to get her attention and reward her bravery.
- Bring a favorite toy to keep her attention and keep the experience positive.
- And remember, give her a lot of praise.

Keeping Your Cat Close to Home: You want to keep your cat's territory close to home so you can be sure his new environment is comfortable and safe. A cat's territory is marked. He knows what's going on in the area, and the sounds and scents are familiar. He finds the choice spots for quiet time and playtime. His territory can vary in size—there are no boundary lines. It depends on the individual cat's sense of adventure, the food and water supply, and protection from the elements.

A male tiger in the wild can range from twenty to over three hundred square miles. A healthy tomcat can control over forty acres. But cats can live quite happily in very small territories if given the proper stimulation and nourishment. As long as an area gives him enough space to play in, nap, and sun himself and he does not feel threatened in any way, your cat will be content. You want your cat's territory to be close to home so he can hear when you call and is always within your safe reach. You can have difficulties, though, keeping your cat's territory close to home in these circumstances:

- Your cat is a female in heat.
- Your male cat is fighting with another male who is on the edge of his territory.
- Your cat is looking for more or better-quality food—at a neighbor's house down the street or at a restaurant in the distance.
- Your cat's hunting and stalking patterns start making him drift farther away.
- Lack of attention at home is causing her to look for outside stimulation.

To keep your cat's territory closer to home you need to make him as content and relaxed with his nearby environment as possible. Follow these guidelines:

- Feed your cat a healthy diet and make sure he always has a supply of fresh water on hand in his indoor *and* outdoor areas.
- If you live in a region where your cat's outside water supply can freeze, provide an electrically heated pet bowl.

- Give him plenty of attention so he won't need to look elsewhere for stimulation.
- Establish routine playtime and activity sessions to give him a chance to practice his hunting skills. This way, he will not need to wander too far from home.
- Teach your cat dinnertime signals so he will get in the habit of coming home for his daily food intake.

I remember Kali's, Serang's, and Rakon's first days at Tiger Island. They were five months old and confident in what they had seen so far in life—but Tiger Island was very new and large for them. Each step taken was slow and careful. They were very jumpy and always looking behind them. All three youngsters were chuffing, trying to reassure themselves. After a few steps of exploring they discovered what they were looking for . . . bushes, and lots of them! For the next two hours the threesome did all its investigating under the protection of cover. Rakon, showing signs of his confident personality, would venture forth every now and then but was quick to head for cover at the slightest noise. Over the course of the week each cub gained confidence, and Serang was the last to come out from cover. Then when it came time to go back home to their dens, nobody wanted to go—it turned into a game of hide-and-seek under the bushes.

When you introduce your cat into a new environment, remember that this new world will be scary and he may react strangely. When you first carry him outside, be careful of loud noises or sudden movements that can scare him into a scratching frenzy. Put a thick towel over your shoulder as a protective measure. And be aware that this is not a time to carry him next to your face. Take your time and give him a lot of encouragement.

IN AND OUT AND IN AND OUT . . .

A cat door is necessary when you have a creature who has a schedule of her own and does not want to rely on you every time she has needs. It frees you from having to play doorman and enables you to carry on with your own life and schedules. A cat door is a flap-covered opening in an outside door large enough to allow a cat to pass comfortably through. If you decide to install a cat door, here are some pointers:

- Pet stores supply plastic or rubber flaps that can be fitted to the hole you cut in your door.
- Be sure to set the flap and hole no higher than two inches from the base of your door so your cat will step rather than jump through it.
- Choose a flap that either opens both ways or allows your cat to enter the house but not leave, according to your needs.
- Buy a flap that has a locking device to ward off unwelcome visitors or make one yourself by affixing a small bolt on the inside of the flap so you can lock it whenever you need to. You will need this particularly to keep back the local toms if your female is in heat.
- To avoid excess draft, purchase a flap with a magnetic strip along the side so the flap will stay secured except when your cat pushes on it.

  To teach your cat to use her new door:
- With the flap set open, place your cat by the door and let her investigate thoroughly.

- While the flap is still open (make sure it is secure and there is no risk of its dropping down on her), go outside and call your cat, offering her favorite treat as her reward.
- Then release the flap so it is down and call your cat, using a treat on the other side to encourage her. Help her push the flap open if she needs it.
- Repeat this many times, and soon your cat will be using her new door with ease.

If putting a cat door into your outside door is not practical, then the two of you can establish your own ritual for her coming in and going out of the house. Soon she will learn how to tell you to let her in, but you too can have control over what form this takes. One favorite method cats use is "bang-bang" against a door. This works particularly well if the door is a little loose and the cat can get momentum behind the bang. Unfortunately, this can become a habit that she does with all doors. It is not one of our favorite feline behaviors. If this cute little habit occurs with your furry friend:

- Ignore her during this time. Any reaction can reinforce the behavior.
- Give her an alternate method of getting in and out of the house.
- Put a doorstop in the interior doors.

A good alternate route for your cat to get in and out is a window. Pick a window that is easy to open and close, easy for your cat to access, in a location that you can easily see, and safe from heights to prevent a dangerous fall. If you decide window access is the best method for your cat to come in and go outside:

- Place her on the windowsill of the open window you have chosen.
- Next, set her on the ground and call her up to the window. (Putting a tasty treat on the sill will help.)

- Give her a lot of praise for a job well done.
- Repeat these steps over and over again.
- Then repeat the same sequence from the *outside* of the window.
- Keep the window open during the day when you're home.

Some pet shops and catalogues even supply cat doors that will fit into windows.

By consistently following what you have taught her, your cat soon will be using the window as her doorway. If you do not feel comfortable leaving your window open during the day, then she will teach you that her presence at the window is her signal to be let out or in. Afterward she will usually thank you with a quick body rub against your leg.

Once we had a window that was well suited as a cat doorway but too high for Mole and Penguin to jump up to. We built them a ladder—a strip of plywood with wood pieces for footing. We also enlarged the windowsill to make a more comfortable resting spot. The window became their favorite hangout. It gave them a high vantage point and also a wonderful sunning spot. This large window made it easy for us to see when they needed to come in and was a great place for them to wait in case we weren't on our toes!

COLD WEATHER CONCERNS

Diet: Outdoor cats and outdoor/indoor cats may require 50 percent more energy to maintain normal body temperature in colder weather. These pets should be served extra portions of high-quality cat food. You must always see that there is fresh drinking water available. Check your cat's water bowl often to make sure the water is not frozen. There are many brands of electrically heated pet bowls.

Antifreeze: Cats are attracted to antifreeze because of its sweet taste, but the ethylene glycol that antifreeze contains is deadly if swallowed. To prevent accidental poisoning, store antifreeze in an inaccessible area. Watch out for areas where a car dripping antifreeze may be parked.

Outdoor Housing: You want to be sure your cat always has an optional refuge to go to when the weather turns inclement. This will help prevent him from getting frostbite. Take a strong box and put it in a sheltered area like a porch, garage, or barn. Put a clean towel or rug inside of it and check regularly to make sure it is dry and clean.

Sore Paws: Snow, ice, road salt, and mud can irritate your cat's paws. Clean the paws regularly and check the pads for cuts.

Fan Belts: Because cats seek warmth under car hoods, more cats are injured by fan belts during cold weather months than at other times of the year. *Always* honk your horn and check under your hood before starting your car.

IN THE COLD REGIONS WHERE
THE SNOW LEOPARD LIVES, HE
WRAPS HIS LARGE BUSHY TAIL
AROUND HIS FACE TO KEEP
WARM.

●

LET THEM KNOW IT'S SUPPERTIME!

When your cat is outside, you want her to associate a signal you teach her with food. There are several methods that are very effective for letting her know it's dinnertime, even when she is outside on the prowl.

The dinner call is important with an indoor/outdoor cat; it tells her that you want her to come inside, with her tasty meal as her reward. Consistent dinner calls will condition your cat so that you can have her on a schedule. If you always choose the same time, your cat will learn to come without being called. (The tigers could never get the hang of daylight saving time.) But you don't want your cat always anticipating the same time to come on the basis of the time rather than a call from you. You will want her to come when you call her no matter what the time is. So you need to establish a call she will respond to at any time of the day, a call that is *positive* and keeps her close to home. To establish a strong food call signal:

- Use a signal, like whistling, that is loud, easy to make, and will carry to all corners of her territory. Judy always calls our cats in because she has perfected the shrillest whistle using her fingers. If you can't or don't want to whistle, a bell, yelling "chow time," or clapping your hands will work. Make sure this signal is not startling or abrasive to your cat. It should be positive for her.

- Once you establish what the signal is going to be, do not change it and use it consistently every time. This signal should be easy for your cat to identify and different from any other that you use. Do not use her name as the signal, since you will use that as a separate signal to get her attention in general.

- When you start, give your dinner call signal in the presence of your cat, then immediately give her food and praise her.

- Next, give the signal with her near but where she can't see you. The moment she responds, reinforce the behavior immediately with food and praise. If she does not respond right away, then move closer to her, still staying out of sight. If she still does not respond, then go back to calling her right in front of her, reinforcing her until she responds consistently.

- Whenever your cat does not respond during a teaching session, back up a step and repeat it until she is consistent. It's usually not the cat's fault, but rather that you are in too much of a hurry.

- Once your cat consistently responds to your dinnertime call, you can increase the distance between you.

- Once dinnertime calls are well established, vary the times so your cat does not learn to anticipate the time, but rather learns to respond to the call. Do this at different times twice a day for both meals.
- Your cat's association of this signal with food will make it a very useful signal, but do not overfeed her. This is not only bad for her health but will also reduce the effectiveness of dinnertime calls as well as your cat's response to them.
- The dinnertime call should never be given without a food reward. You want to keep the association with food strong. If you need to get your cat to come in and do not want her to have a meal at that time, then a tasty treat will do. Remember, you and all your family members need to keep track of the treats given to your cat so she is not overfed.
- If you are feeding your cat twice a day and you feel she is wandering too far away in between meals, then split the portions up into equal parts and feed her three or four times a day.
- If your cat is straying too far from home, keep changing the time she is fed. This will keep her guessing, and force her to stick closer to home—she won't want to miss a meal!

COMMON CAT FIGHTS: WHAT TO DO ABOUT THEM

As long as there are cats, there will be cat fights. There are ways you can avoid cat fights and the best way is to spay your female (so as not to attract males) and to neuter your male.

Even if your cat is spayed or neutered, cat fights can happen. It is natural for cats to protect their territory from an unwelcome intruder. Spayed or neutered cats tend to stay closer to home and their territory is usually much smaller. Still, if your cat is new to the area, he will try to develop his territory—and there may be a cat who already has a claim on it.

You need to keep a close eye on your cat when he is establishing his territory or an unwelcome intruder is invading his area. You can help him by these methods:

MOST CAT FIGHTS OCCUR DURING THE MOST ACTIVE MATING PERIOD—THE SPRING. MATING LESSENS IN THE SUMMER AND STARTS UP AGAIN IN THE AUTUMN AND WINTER.

●

- When you see a confrontation coming, try to scare the other cat away by making noise of some kind. A hose or squirt gun will also make the other cat feel unwelcome and give your cat a better chance to establish his territory. *Do not* throw anything at the other cat; you are only trying to encourage him to find a new territory, not hurt him.

- If noise or water does not discourage the other cat from your cat's territory, then ask around and find out who the owner of the cat is. When you find the owner, ask about the history of the cat. If that cat has not been spayed or neutered, recommend it as politely as possible—perhaps relating some of the dreadful statistics in Chapter 10 might help persuade him.

- If your cat has been "fixed" and is still fighting or even instigating the fights, then schedule more playtime and activity into his day. This will help get rid of his excess energy.

- If fighting persists, use dinnertime calls and keep him in for the entire night, since the majority of cat fights happen at night. If you suspect the problem is a female in heat attracting other males, keep your cat in at night for a few nights. If you suspect your cat has been fighting, he needs to be checked frequently for injuries and scratches. Even a healed wound can develop into an abscess.

Occasionally the tigers fight with each other. It's usually sparked by play that gets too rough and results in a fight. If Kali, with her playful nature, picks on Rakon too much, he retaliates and chases her around. If he's really fed up with her, we break it up quickly, since this kind of behavior can get the cats hurt.

We break up tiger fights by distracting the tigers, the same way you do with your cat (but much more carefully). Because the tigers are well conditioned and know to listen to us, we are able to divert their attention from one another. We usually call the name of the aggressor, in a stern voice, letting him know this behavior is not tolerated. **Remember to be especially careful around any animals when they are fighting to avoid being injured yourself.**

RODENTS AS GIFTS

A YOUNG AND CRAFTY CAT IN GOOD PHYSICAL AND PSYCHOLOGICAL CONDITION CAN BRING IN A "BOUNTY" OF AT LEAST 1,000 MICE IN A YEAR.

●

The ultimate compliment your cat can give you is to bring home a gift. In many cases the owner does not see it that way and in fact may protest quite loudly. But you as the mother figure need to let your pet know what a wonderful hunter she is. A shocked or displeased reaction from you can be very baffling to your cat. In the wild, it's the mother who helps her offspring develop their hunting skills. Mother tigers bring down the prey and only cripple it to give their young the opportunity for their first kill.

If you want your cat to help control the mouse population in your area, then praise and thank her when she brings home a rodent. This reinforcing response will encourage her to do it again. If this is not behavior you want your cat to continue, then ignore her when she brings you her prey. Simply pick up the dead mouse and dispose of it. Try not to react and never yell or scold your cat. By your ignoring her behavior, your cat will decide not to bother next time.

Many cats offer also other gifts like socks or underwear. This occurs mostly with indoor cats, but any cat can get into this habit. This means your cat is working hard to please you, the mother figure, and she should be given a lot of reinforcement for this behavior. You may not wish for the gift to be an article of your clothing (especially when you look in your sock drawer and there are no matching pairs), so you need to introduce your cat to a toy she can use instead. Once she consistently brings you the toy as a gift, you can turn this into a fun game for her by hiding the toy to see if she can find and return it to you.

The more elaborate the game of hide and seek that evolves with this favored item, the more proud your cat will be when she returns it to you.

BIRDS

As Harriet Beecher Stowe wrote about her cat, "She could never be made to comprehend the great difference between fur and feathers, nor see why her mistress should gravely reprove her when she brought in a bird, and warmly commend when she brought in a mouse." For the outside cat, stalking birds is a fact of life. Birds are everywhere and very stimulating to the hunting cat. So curbing the hunting instincts a cat has for a bird can be difficult. I've always felt sorry for the unsuspecting duck who flew into the pool of Tiger Island.

If your cat has overdeveloped his hunting skills and has been taking advantage of the bird population, you need to give him some other options:

- Increase his playtime schedule, using many toys on a string or rope, to burn off excess energy and allow him to practice his hunting skills in a less damaging way.
- Make sure his diet is sufficient enough to satisfy his appetite. His motivation for hunting could be food-related, though many cats hunt strictly for the excitement.
- If you have a bird feeder, you could be increasing the odds of his success. If you want to continue to observe the beauty that your bird feeder draws, then make sure it is out of the reach of your cat and in a wide-open area. Remember that cats take cover when hunting.

CAT ON A LEASH

If your surrounding area isn't quite safe enough for your cat to go out on her own, you can condition her to walk on a leash. Dogs aren't the only pets who can be walked on a leash. Your cat can grow to love a walk on a leash too—it just takes a commitment of time on your part.

Each one of the tigers at Tiger Island is leash-trained. This is essential for moving them from one location to another in a controlled and safe manner. Once they're trained, there's nothing to it. We simply walk over, put a sturdy chain around their necks and we're off! This is great for them since we can take them on walks outside of their usual environments. This keeps them stimulated with new experiences for their senses and lots of fun spots to investigate.

Walking your cat on a leash is great exercise—for both of you. It is especially welcome if you live in an apartment or a big city and have a park or nice area to walk in. The training will take enormous patience and dedication, but hang in there. Once she gets it, it will stick with her. It's easiest to start with a younger cat, although with more persistence on your part an adult cat will be able to master walking on a leash as well. You need to go out on a daily basis, for it is through repetition and making the walk a positive experience that your cat will learn to walk on a leash—and enjoy it.

People ask what keeps a tiger, who weighs almost three times my weight, from running away on a leash? My response is always the same: training with consistency. Even a large dog is difficult to stop if he decides to take off—I once saw a lady being dragged down the beach by her Saint Bernard. But with repetition and training, a dog learns to walk on a leash, realizing that otherwise he doesn't get to go out. Because we have established a mother role figure with the tigers, they listen to us and it doesn't affect their behavior that they outweigh us and are much more powerful.

Rakon is a great walker and has trained me well about when it is time to rest. He starts a series of moans and then plops down by a bush, nibbling at a branch with his large canine teeth. I want to keep his walking experience positive for him, so I listen and let him take frequent rests during our walks. Because we have been able to keep walking on a leash such a positive experience for Rakon, we can take him anywhere.

It is very important that your cat trusts you and has confidence in you. She will look to you for guidance in this new situation, and you need to be reassuring. Here's how to teach your cat to walk on a leash:

- You should buy either a leash or a harness. A harness is the better choice as there is no chance of your cat's head slipping out of it.
- In the room or area where she is the most comfortable, show her the new leash or harness and let her investigate it.
- Once she has completely investigated it and is relaxed, slowly place it on her. At this point just let her sit or walk around on her own with it, building confidence and getting used to the leash. She should never wear a leash without your strict supervision—it could get stuck around her neck and choke her.
- After she is relaxed and comfortable with a leash on her neck, while still in familiar surroundings, gently take the other end and take some short steps with her. If she does not resist and steps along with you, give her a lot of praise. If she does resist, then praise her for wearing the leash and allowing you to hold the other end until she has gotten used to it and you can proceed to the walking stage.

  Once she is relaxed wearing a leash with you on the other end in familiar surroundings, it is time to go outside. Pick a quiet environment and a small area.
- Keep the first walk very short, with many stops to allow your cat to investigate all the new sights, smells, and sounds. Offer praise and treats.
- Expect a lot of zigzagging in the beginning rather than a constant straight walk. There will probably be several short spurts of walking with investigating, rolling around, or chewing on grass in between.

- Be alert for any noise or object that could startle your cat into trying to flee.

The main reason people have difficulties leash-training their cat is because they expect too much too soon. Your patience is essential. Only *slowly* increase the distance and duration of your walk. It will be up to you to keep this a positive experience for your cat. Make sure to investigate the walking area before taking her there. And take the time to make your cat comfortable with known routes before investigating new routes.

BATHING: A FUN DAY OUT MAY RESULT IN THE NEED OF A BATH!

If your cat is an indoor/outdoor cat, he will probably need a bath from time to time. Many cats never really need a bath, especially if you follow the grooming ritual regularly during quiet time and give your cat a nutritionally balanced diet that keeps his coat in good shape. But some long-haired cats and indoor/outdoor cats (especially ones who have just chased a bug into a puddle of greasy oil) need an occasional bath. There are also medical reasons why your cat may need a bath, such as flea infestation. A tiger will naturally get his bath by taking a cool dip in a river or pool. It's a tiger's nature to like water. Most house cats tend to avoid direct contact with water, which makes your job more challenging. Here's how to go about bathing a cat:

- On hand you should have a few absorbent towels, a plastic cup or spray hose attachment for rinsing, some cotton balls, and a quiet hair dryer.

- Don't use any shampoos with chemicals or conditioners unless your vet has prescribed a certain soap for fleas. A pure castile soap works well. There are also chemical-free and environmentally safe pet shampoos.
- Make sure the room you have picked for your cat's bath is warm and draft-free. Try either the kitchen sink or a large plastic tub placed in your bathtub. Place a rubber bath mat in the bottom of the sink or tub to give him something to grip onto for better balance. Fill up the tub or sink with two to four inches of water.
- Give the water the "baby bath" test and make sure it isn't too warm. Also check the shampoo, making sure it isn't too cold.
- During a quiet time make sure you've completely brushed your cat out and there are no tangles or mats left on his coat before bathing him.
- Water can be terrifying to your cat, especially the first time, so wait until he is calm and relaxed, like right after a good quiet time session together. Reinforce his calmness using a soothing voice.
- Gently place a cotton ball in each of the cat's ears to prevent water from going into his ear canals.
- Place him in the water, slowly, hind legs first so he can see you for reassurance.
- Keep a secure grip on him and slowly start wetting his back and sides, working your way up. Always do the head *after* the rest of the body is lathered.
- Once his fur is completely wet, apply the shampoo. Be very careful not to get any soap or water in your cat's eyes, ears, or mouth. Massage the lather all the way in to his skin. After the body is done, wet his head and lather it too. Watch to see if he is getting chilled,

and if he is, pour some warm water on him. If he is really greasy, you may have to give him two lathers.

- Now it's time for the "rinse cycle." A spray hose attachment works great, but make sure you keep the water pressure down so you don't scare him. If you don't have the attachment, then turn running water on and find a lukewarm temperature while still keeping that secure grip him. Then use a plastic cup to rinse him, and rinse until the water is completely clear.

- Take a towel and blot-dry your cat. Try using warm (not hot) towels fresh out of the dryer, changing them as they get wet. Make sure your cat is staying warm.

- Now you can use a hairdryer on a *low* setting. Keep it moving and never leave it on one part of him or you risk burning him. If your cat has long hair, comb and brush it while you finish drying him to keep it from tangling.

As soon as this not-always-so-pleasant experience for your cat is over, be sure to let him know how beautiful he looks and give him an abundance of praise—and his favorite tidbit!

THE BRITISH TURKISH HOUSE CAT
ENJOYS PLAYING IN WATER.
●

cat on the move

*Everything that moves serves to interest and
amuse a cat.*
F. A. Paradis de Moncrif

SHOULD CATS TRAVEL?

By introducing your cat to traveling at an early age, you will be able to introduce her to all sorts of new experiences away from home. Some people think only dogs like to go on trips and car rides, but this isn't so. Cats can make great traveling pals and it can enhance your relationship by allowing you to spend time together when you otherwise might have to be apart. Letting your cat experience new surroundings is stimulating for her and good for her health. Many are the benefits to having your cat accustomed to traveling.

There are times, too, when it is essential for your cat to travel, as when you take her to the vet or you move, and conditioning her to traveling reduces stress for these mandatory trips.

The sooner you begin to train your cat to be a traveler, the better. Some cats will adjust immediately, enjoy this new experience, and eagerly look forward to their next trip. Others may be nervous and frightened in the beginning and require your understanding and patience. You need to build your cat's confidence to travel.

THROUGHOUT HISTORY THE CAT HAS TRAVELED. WITH THE GOOD COMMERCIAL RELATIONS EXISTING BETWEEN ASIA AND EUROPE, CATS WERE TRADED FOR SILK IN CHINA. CATS WERE INTRODUCED TO JAPAN FROM CHINA IN THE MIDDLE AGES. ACCORDING TO LEGEND, THE FIRST CATS APPEARED IN JAPAN IN THE YEAR 999 IN THE IMPERIAL PALACE OF KYOTO.

●

If your cat is not an ideal traveler overnight, you will be able to develop her ability to go anywhere in just about any situation through conditioning. The conditioning should start in your house and a good way to start is with "travel box training." Once you train your cat to use the box, you and she will be able to share almost any mode of transportation.

TRAVEL BOX CONDITIONING

A travel box is as much of a necessity as a food bowl or cat toys. Of the many models available, the one you choose for your cat should be comfortable and

- Made of a strong material, built to withstand impact.
- Well ventilated.
- Leakproof.
- Easy to clean.
- Simple to dismantle for easy storage.
- Able to double as a kitty bed or cozy home.
- Suitable for airline transportation. Be sure to check the dimensions specified by airlines for carry-on cases.
- Not too large, since this could result in your cat being bounced around.
- Big enough for him to stand up and turn around.

Conditioning your cat to be comfortable in a travel box is not a difficult task and should be started as soon as your cat is comfortable with his home environment. The time will come when your cat must go somewhere, for example, to the vet, and if you neglect this important training, you will have to force him into his box which will make it a negative experience for him—and even more difficult the next time. The main mistake people make when training their cats to use a travel box is not allowing enough time for the training. So when you have found and purchased the right travel box for your cat, follow these steps:

- Start in your cat's most comfortable location and let your cat investigate his travel box. Leave the door open and make the box inviting by placing bedding inside; tempt your cat to use it as a bed.
- Once your cat is completely relaxed about the box, gently place him inside it, and leave the door open. Keep him inside for a very short time. Repeat this, slowly increasing the time spent inside as long as he isn't resisting.
- When your cat is completely comfortable inside the box, shut the door for a short period of time. Increase the time inside according to your cat's confidence. Be very careful that his whiskers, tail, or paws aren't sticking out.
- Next, start to carry the box with the cat inside around your house until he gets used to being carried around to different locations. Offer plenty of praise and don't push him if he isn't comfortable. You want to make this as positive an experience as possible.
- Finally, move the box outside and eventually to your car or the transportation you will be using. It is very important that you do not rush any of these steps. Your cat should be very relaxed in his box in the car before a journey is attempted. And before going on a long journey, try out a lot of short trips of varying durations.
- As with all conditioning, always give your cat a lot of reinforcement to make staying in the traveling box a positive experience. And remember to use food rewards.

Each of the tigers is taught to travel when still very young. In the beginning we use a travel box similar to the one you will use, but this doesn't last long because tiger cubs gain weight fast—about twenty-five pounds in three months. One of my favorite tales is an example of what can occur if the cage or area you're trying to get your cat into has become associated with a negative experience.

I was working with a young lion by the name of Sennetty, conditioning him to go into his extra-large traveling cage so he could be transported when needed. All of the other big cats at this time would stay in their travel cages with the gates shut for long periods of time. Well, with Sennetty it was a different story. In the beginning stages of conditioning he had misjudged the doorway and banged the top of his large head on the gate frame. That was all it took to make entering the traveling cage a negative experience for this young lion. The bump on the head didn't hurt him, it just gave him a good scare—enough to make travel cage training a real challenge. We had to work the entire day trying to get him to go in, taking many time-outs, giving him a lot of praise when he got near the cage. To make matters worse, it started to rain and Sennetty started his loud OWWOO, the way upset lions do. I tried putting his favorite treat (a twenty-pound slab of horsemeat) inside of the cage. I tried everything. But he was not going in no matter how tempting we made it. After an entire day of failed attempts at getting Sennetty to put even a paw into his travel cage, we went home feeling quite defeated.

One of the other trainers, Pat, came in that night to check on everything and see if Sennetty had gone into his travel cage. He read the records that we keep on all the tigers and lions and found that he had not, in spite of all of my efforts. Pat always drives around in his truck with his house cat, Sumo. Sumo was patiently waiting for him when Pat came up with a great idea for getting Sennetty into his travel cage. He quickly ran to his truck and grabbed Sumo.

Sennetty was near the entrance of the travel cage, and Pat went around behind the cage to a spot where the lion couldn't reach but could see. Pat held up Sumo the house cat and displayed him to Sennetty the lion who immediately flew into the cage with wide-eyed excitement, completely forgetting the fear he had displayed all day. From that day forward we never had a problem getting Sennetty into his travel cage.

Lions and tigers like cats . . . as a tasty meal. When Sennetty saw Sumo, all of his hunting instincts came out, making him charge in the direction of potential prey. Once he was in, Pat put Sumo back into the safety of his truck and rewarded Sennetty with praise and meat. Sennetty's fears of the cage were over, and in fact, he now associates the cage with an exciting and positive experience (stalking).

THANKS TO HIS EQUILIBRIUM, THE CAT, UNLIKE THE DOG, DOESN'T SUFFER FROM CAR- OR SEASICKNESS.

●

All cats are instinctual, and many times if you think like a cat and put yourself in his paws, you will be able to solve any problems that may occur.

CAR TRAVEL

Car travel can be very stimulating for your cat, especially in the beginning stages. After several trips, some of the excitement can wear off and your cat will simply enjoy your company and being out.

The easiest and safest way to transport your cat in a car is in her travel box; otherwise she may roam around the car and possibly interfere with your driving. Also this precludes the possibility of her darting outside the car every time you open the door or getting into a position, such as under a pedal, where she could harm herself.

However, there are many people who prefer to have their cats riding in the car not in a box. If you are one of these, it is very important that your cat is conditioned to stay in "her spot" and not move around. Here's how to make a safe and seasoned car rider out of your cat:

- Start after your cat is relaxed with her home environment and is already leash-trained, if possible.
- Carry or take your cat out to the car by leash and gently place her in the car, letting her completely investigate the entire interior.
- Offer her a food reward when she relaxes.
- Place her on the passenger seat and again offer her a food reward for staying there.
- Never let her on your lap, and use a firm "no" if she goes anywhere near the driver's seat or controls.
- Work on giving her positive reinforcement for being relaxed and staying in the passenger seat.
- Once she consistently stays in the passenger seat and is comfortable in the car, start the engine. Some cats will be frightened of this new sound, so wait for her to calm down. If she has moved, wait for

her to go back to her seat. Then offer a treat and a lot of praise.

- When you are confident that your cat is comfortable with the sounds of the engine starting and running, take her on a short drive around the block. The most important focus should be for her to stay on her own seat and not to move toward you while you are driving.

- Once she has learned this, practice opening and closing your door, with you both in and outside of the car. Use a food reward to reinforce your cat's staying in her seat and not slipping out the door when opened. This is of the utmost importance.

- Whenever your cat travels by car, make sure she has *ample ventilation*. Also, carry a container of water and a water dish to make sure that water is available any time she may need it.

- Slowly increase the distance and durations of your trips. If your cat gets restless and starts moving around, stop the car, say "no" and firmly pull her back on her leash or grab her and place her back in her spot. Give her a reward when she stays.

- Before you go on a long trip with your cat, talk to your vet and make sure she has a clean bill of health. Do not travel with a cat who is ill, pregnant, very nervous, weak, or in heat (unless it's an emergency trip to the vet).

Traveling properly, like any other behavior, must be maintained and therefore practiced frequently.

Because the tigers of Tiger Island are so conditioned to travel, we have been able to introduce them to many new environments and situations. We've also experimented with several different modes of transport for them. We make "tiger taxis" out of shopping carts for them when they are younger. We push them all over, letting them see new surroundings. They love to stand up with their paws on the sides of the carts watching the world go by. When they get bigger we put them in a golf cart with their own customized backseat. This is definitely Rakon's favorite mode of transportation and at times it can be a real pain to get him out of the cart.

The tigers also enjoy traveling by boat, and we often take them for a leisurely ride on the lake near Tiger Island. Once Kehei was showing that he enjoyed the boat ride by leisurely dragging his tail and paws in the water. He saw something dart in the water (probably only a reflection) and reacted by leaping into the lake. I think he was more surprised than I was as his sleek body hit the water. I decided to let him swim ashore by keeping a loose grip on his chain and slowly moving the boat to his pace. He was a bit tired after reaching shore, shook himself off, and settled down for a long snooze. Kehei now knows to look before he leaps!

For longer trips, a van turns into a large "tiger traveling box" with good ventilation and plenty of straw for bedding. I remember Rakon and Serang's first trip to the mountains, where they were able to experience snow! At first they were timid and kept slipping on the ice. Then they quickly adjusted by taking little steps. When they got to the deep snow, where they sunk to their chests, all I could see were two great balls of orange and white.

TRAVELING BY AIR

If you want to travel by air with your cat, check the rules about bringing him aboard. Some airlines allow this if your travel box qualifies. This is the least stressful way for your cat to travel by air. You will still be able to talk to him and check his mood and needs.

It is much more stressful for your cat to travel as air cargo. It can also be dangerous—something could fall on the travel box or your care instructions may not be followed. If you cannot take your cat aboard, consider whether it might be better to leave him behind in trusted hands rather than submit him to the stress and risks of flying alone in cargo. If you do decide to send your cat air cargo, follow the USDA rules. (The USDA regulates the transportation of animals.) To prepare your travel box for airline usage, you need to

- Mark clearly on the sides, in letters not smaller than one inch, "LIVE ANIMAL."
- Mark your pet's name and phone number of the final destination.
- Mark arrows indicating the upright position of the traveling box.
- Put clean bedding such as straw, shredded paper, a blanket, or cushion in the box.
- Make sure water will be available at least every twelve hours after the trip is initiated.
- Display feeding and watering instructions even if your pet is to receive no food or water.
- Contact the airline if you have any questions.

(Many trains and most buses do not allow animals on board. Also, certain states require a health certificate and proof of rabies vaccinations.)

TRIP TO THE VETERINARIAN

Your cat will need to go to the vet for a general checkup and to update vaccinations once a year. You of course also need to take your cat to the vet if an accident occurs, if she gets sick, or if she is injured. Going to the vet can be a very stressful experience for your pet. If she is not comfortable with traveling, you are adding stress to stress. Her stress level during trips to the vet will be reduced if she has been conditioned to traveling. But there is a way to reduce her stress level for an actual visit to the vet:

- When you are out taking a drive with your cat, stop in at the vet's office and take her inside.
- Bring treats and reward her inside the waiting room, making this a positive area.
- If your vet is willing and has a few minutes, have him come out and pat your cat and talk to her. Most vets are happy to do this, for they know in the long run it will help them when treating the animal.

We learned this from Baghdad, who has a fierce dislike for her vet. When she was a cub growing up, she would see the vet only when she had to get a vaccination or treatment. Of course, this became very negative for Bag and she would hiss and snarl whenever she saw the vet, even from a distance. We started taking the younger tigers (too much water under the bridge for Bag) to see the vet on a regular basis, just for friendly visits. After doing this for a long time, we discovered that the vet was able to get closer to the tigers for treatment, and they responded positively to her.

BECAUSE OF STRICT QUARANTINE LAWS, THE FOLLOWING PLACES ARE RABIES-FREE:
- GREAT BRITAIN
- AUSTRALIA
- NEW ZEALAND
- HAWAII

VET TALK

You are your vet's best source on your cat's better health since you spend the most time with him and know what his normal behavior is. Try to establish a communicative relationship with your vet. Remember, you understand your cat's conduct better than anyone, and the information you provide can be very helpful when a problem arises.

This is a list of infections and recommended vaccines for them that all cat owners should know about and discuss with their vet. A good rule of thumb if you adopt a stray cat or "inherit" one: Assume the cat will need all the shots, even if he is an adult.

1. **Feline Distemper** To prevent respiratory infections, a vaccination sometimes called a 4-1 shot should be given at the age of 8 weeks, then repeated two more times at 4 weeks apart. Then a booster should be given annually.

2. **Feline Leukemia** This is a very deadly disease that is passed from cat to cat through blood and saliva. With the exception of automobile accidents, it's the number one killer of cats each year. In the last five years a vaccine has been developed that now gives owners a chance to fight this disease. Kittens at 12 weeks should be vaccinated initially with 2 doses, 2 weeks apart, and a booster should be given annually.

3. **Rabies** With the increasing number of cats in the United States, this disease is now carried by more cats than dogs. Kittens at the age of 16 weeks should be vaccinated, and a booster given annually.

CHAPTER 9

Your cat in new situations

> Watch a cat when it enters the room for the first
> time. It searches and smells about, it is not quiet for
> a moment, it trusts nothing until it has examined
> and made acquaintance with everything.
> —Jean-Jacques Rousseau

CHANGING ROUTINE

You know he's there, you see a glimmer of light reflecting from his eyes, and no matter what your tactic, he won't budge from under the couch. "C'mon, after a nice ride in the car we'll be at our big new house with lots of birds around it." or "Hey, what's so terrifying? It's just a little golden retriever puppy wanting to play!" or "Aunt Wendi and Uncle Joe and their kids will play with you all week while they're here." The name "scaredy-cat" sure fits him now.

Why are some cats outgoing and aggressive and others shy and timid in new situations? It is all a matter of whether the cat has been socialized and adjusted to different places, people, sounds, sights, and situations from the beginning. For example, a cat who has been raised in a house full of people will react differently from a cat who has had little exposure to people. Introducing your cat to people as soon as possible will allow him to be comfortable around strangers. And although cats are solitary animals, they can easily be conditioned

177

CATS CAN PERCEIVE A DECREASE
IN ATMOSPHERIC PRESSURE,
MINOR EARTH TREMORS, THE
SMELL OF DISTANT RAIN AND
CHANGES IN A DAILY ROUTINE,
SUCH AS WHEN AN OWNER IS
ABOUT TO LEAVE HIS HOUSE.
•

to interact and have a compatible rapport with not only people but other kinds of animals as well.

Cats are instinctual and react to anything out of the ordinary— it's not their nature to try to figure a situation out. For every different situation they react differently. Cats are unpredictable in new situations: They may run away, freeze up, or get excited.

Younger cats and cats who have been socialized will better handle changes. But any change in routine can be overwhelming for a cat, especially one who has been used to a set routine. And it is up to you, the mother figure, to make it easier for him to accept and adjust to the change. Here are the basic rules for introducing *any* change into your cat's routine:

- Before the change, offer *extra* attention and praise to him, making him feel secure.
- Think and plan ahead.
- Be alert and watch for behavioral changes.
- No matter what the change is, try keeping other routines the same, such as playtime and quiet time routines.
- Stay calm. If you are nervous, your cat will sense this and become anxious. If your cat overreacts to the new situation, for his sake, you need to stay composed.

MOVING

Moving to another house is a major cause of stress for us humans. It is worse for our cats! Cats detest change and moving can be an overload on their senses. The best thing you can do to console your cat when moving is to offer her as much attention as possible and take notice of any change in her behavior before it turns into a bad habit.

At the beginning of a serious change such as moving, cats can be overreactive and sensitive to their surroundings. You may notice a temperament change, with your cat being distant, unsettled, or destructive. Give her some time and do not expect or demand your cat to adjust quickly to her new home. Your impatience will only add to her stress. In time, new patterns will develop and the two of you can choose new areas for feeding, quiet time, and playtime. And soon your cat will soon stake out her new territory and investigate new favorite spots to sun and nap.

However, there are some ways to help your cat adjust to her new home:

- Moving will be an easier transition for your cat if she is already travel-box-trained. The box will be a familiar territory for her, and make her feel more secure.
- Keep her indoors for two days before the move and give her a lot of attention. Cats have been known to sense that something is going on and have run away *before* the move.
- As always, be sure your cat has a safety collar and tags, or some kind of identification before you start your move.
- Feed her at least four hours before the journey.
- When you first arrive in your new house, put your cat in a room with furniture that is familiar to her and may comfort her.
- Go through the "household hazards" checklist (pages 120 to 121).
- Gradually introduce her to new rooms and unfamiliar territory.
- Maintain as much of a normal schedule as possible.
- Provide many play sessions and shared quiet times and give her as much attention as possible.

- Be patient with her and take note of her disposition and stress levels even during the business of unpacking.
- If your cat is an inside/outside cat, make sure she is *completely* adjusted and familiar with her new home before you let her out.

In your house, try what Hayward, California, animal services officer Mary Brown suggests: Prepare for disaster (fire, flood, etc.) by keeping portable carriers or kennels stocked with extra collars and leashes, tie-out cables, bedding, water, nonperishable pet food, a manual can opener, a first aid kit, vaccination records, and toys.

The older your cat is, the longer it may take for her to adjust to her new environment. But be patient and the two of you will soon be content in any new house.

INTRODUCING YOUR CAT TO A NEW HOUSEHOLD PET

When introducing a new pet to your cat, you must plan ahead to make the change as easy as possible and to ensure the health and safety of both. Here's a checklist to use before you introduce them:

- Does your cat have all his shots and is he in good health?
- Does the new pet have all of his shots and is he in good health?
- Has your cat's behavior been normal lately, or does he seem to be stressed? You don't want to make any changes for your cat if he isn't himself or is overly stressed.
- Is your pet neutered? This is a must for both the newcomer and

DOES A CAT HAVE A HOMING DEVICE? RECENT RESEARCH SUGGESTS THAT THE HOMING ABILITY OF CATS LIES IN A FORM OF BUILT-IN CELESTIAL NAVIGATION, SIMILAR TO THAT EMPLOYED BY BIRDS. THE BRAIN AUTOMATICALLY REGISTERS THE ANGLE OF THE SUN AT CERTAIN TIMES OF THE DAY.

•

your cat. If you bring home a kitten, wait until the proper age (for male cats after they are six to nine months old and for females at six to seven months, after the first heat period) to make the arrangements for neutering. This will also help avoid fights with males.

• If your cat recently has been neutered or received an injury, make sure he has fully recovered before bringing home a new pet.

Once you have gone over this checklist and you're ready for a new pet, remember that the introduction of a new pet, or household member, can be tricky for your cat and needs to be handled delicately. His reactions you want to watch for are: arching his back, a strong hiss, a disappearing act, his litter box routine gone astray, aggressiveness, and general misbehavior of a type not seen before. He will need lots of attention to feel secure with this situation. So too will the new pet. Here are some pointers:

- Give plenty of praise to your cat during the transition period.
- Let someone else bring in the new pet so your cat does not associate you with the new pet.
- Introduce them in a calm manner in a quiet room with no other distractions.
- Give plenty of attention to the new pet too.

These four fundamentals should be followed when you bring home any new pet, but there are also specific helpful procedures to follow for different types of animals.

Cat Meets Cat: You can't predict the reaction two cats are going to have when they meet; it depends on the age, sex, and temperament of both.

Sampson is unique—he loves to play with cubs, which is extremely unusual behavior for a male tiger. In the wild, a male tiger would rarely be seen with cubs; it is the mother's responsibility to teach the cubs to survive. When Mohan and Rakon are around cubs, they are nervous and unsure what to do, and they always end up looking for a quiet spot to escape the youngsters.

We are always very cautious when introducing the cubs to an older cat because of the size difference. The cubs are

usually frightened at first and will hiss and run away. Once they are relaxed around Sampson, he allows them to crawl all over him, tug at his tail, and nibble on his ears. He playfully and gently throws them down and places his giant paw over their miniature bodies. They squirm until they get away, then shake themselves off to get ready to bite back at the large moving target. Cubs have an abundance of energy and tend to bite each other a lot. Sampson is very tolerant and seems to enjoy their shenanigans. Sam is the exception, not the rule.

You should introduce a new cat to your pet slowly and carefully. To him, you're bringing in a new child, friend, and playmate, who is infringing on his territory and, even worse, is competition for his mother's time. He will want to protect his territory and the new cat will be nervous in her new environment and will be ready to protect herself. To introduce the two in a manner that will be less traumatic for them both:

- Keep them completely separated until the new cat has had a chance to settle down and become familiar with her new surroundings and your role as the mother figure has been established.
- Then place the cats in separate rooms connected by a door, which serves as a "buffer": They can smell and investigate each other from under the door.
- Let each investigate the scent of the other on a towel.
- Switch the cats' rooms, letting the new cat become familiar with different surroundings.
- Go slowly. You want to be sure your cat is used to the idea of a stranger and the new cat is relaxed in her new environment.

- When you feel both cats are ready, let them see each other by having someone else bring the new cat into the room where your cat is.
- If your cat is leash-trained, for better control have him on the leash when you introduce him to the new cat.
- Gently set the new cat down, and while supervising carefully, watch for their reactions.
- If they fight, stay calm and direct your attention to your cat, not the new one. Call his name and try to divert his attention. If the fighting gets out of control, place a large pillow between them to break it up. Separate them and give them a long period of time to settle down and relax before you bring them together again.
- Keep the first session short, but allow enough time for them to become well acquainted. Afterward, it is important to reinforce your cat's security with a lot of praise and a tasty treat.
- When you feel they are getting more relaxed around each other, increase their time together, but always make sure they are supervised in the initial stages.
- Only when they have become used to each other and there is no unfriendly behavior can you start leaving them together unsupervised.
- Even after your cats are relaxed together and interacting, wait still longer before you feed them side by side or have them share a litter box. This could be added pressure for them to socialize before they are ready.

  A good alternative buffer to a door is a travel box. If you have already conditioned your new cat to her travel box, having her in it is another good way to introduce them and see what kind of reaction you get. Do this by placing the travel box with the little newcomer inside next to your cat. Let them investigate each other, and take note of your cat's reaction. If your cat is indifferent, then you can put them together sooner. If your cat has an aggressive reaction, then you have to take more time to introduce them.

It is a very difficult task to introduce the tigers, especially the older ones, to a new tiger. Being very large and powerful, they can easily inflict injury on one another quickly, which can even lead to a fatality. In most cases we introduce a new cat only to young cubs or adolescents who are more accepting, playful, and willing to adapt. The first time Sampson was put together with Tara and Jai (two sisters), he was very nervous. He had already become familiar with Tiger Island, but having to meet two of the local residents was a big step. We tried to keep his first encounter under control by starting with one cat at a time—Tara first, then Jai. Sam was very jumpy and would roll on his back in a defensive position if Tara was getting too rough. He would chuff at first, then snarl if Tara got too close. Tara and Jai were kept on leashes the first couple of times to ensure better control. If Sam got too nervous we could have the girls lie down until he relaxed. With our having that control, Sam wouldn't become too defensive and in a short time he felt comfortable and did not overreact to the girls' company.

THE LION IS THE ONLY FELINE THAT USUALLY LIVES IN A FAMILIAL PACK. THIS PACK IS CALLED A "PRIDE" AND HAS ONE LEADER. OTHER FELINE SPECIES WILL COME TOGETHER FOR SHORT PERIODS OF TIME, PRIMARILY TO BREED.

●

Cat Meets Dog: Even though their personalities are different, dogs and cats are able to establish a close relationship under the right circumstances. If you want to introduce a new cat or kitten into your home, and already have a dog, you will want to do so in the *least stressful* manner for the cat.

- First, give the new cat or kitten time to adjust to her new home and "mom."
- Once you are confident that the cat is well adjusted to her new surroundings, introduce her to your dog by bringing your dog in on a leash.
- Check your dog's reaction first. If he is too playful, rough, or excited, then keep him on a leash to maintain tight control.
- Next, check the new cat's response. If she seems too nervous or stressed, then keep the session very short. The cat will feel assured that mom is protecting her and will be able to slowly adjust to that big furry thing.
- When the cat and dog are familiar with each other while your dog is on a leash, and you are sure that both are comfortable, then take your dog off his leash. Don't let your dog chase the cat in a playful manner. This is usually fun only for the dog.
- Keep their activity supervised with the dog under control. The most intimidating factor for most cats is usually the extreme size difference.
- Give lots of attention and praise to the new cat after interacting with your dog—do this without the dog around, since a jealous dog will cause only more stress for the cat.
- Don't forget to reinforce your dog also and offer him a tasty treat for accepting this new member of the family. This will also help keep

him from getting jealous at your having to share your attention with another animal.

- If your dog or cat shows signs of aggression, have "interactive sessions" with your dog on his leash. Praise good behavior and give a "no" for bad. Do this often and have patience.

Cats and dogs together can make a wonderful combination. We have a golden retriever named Kezar that has grown up around his fair share of cats—house cats and tigers. As a young puppy he was first introduced to a litter of tiger cubs. Over the course of a summer he would spend hours playing with the tigers. The afternoons were a frenzy of activity. He had an endless amount of energy that would keep the young tigers in a much more active and stimulating atmosphere. Kezar learned to play in a gentle manner with all cats, and they in return learned quickly that if they played too rough with sharp teeth, they would get a very loud, reprimanding bark.

Because Kezar established such an incredible relationship with the cubs, he and the cubs were invited to be guests on *The Tonight Show (Starring Johnny Carson).* On the trip to L.A., Kezar was the perfect baby-sitter, keeping the cubs entertained in the back of the van. During the show, it was a different story. As is the case with all animals, the unusual happens. Kezar was supposed to mind his young charges onstage, but instead the cubs went every which way, while Kezar investigated the crowd, much to their delight. The segment finished with the spotlight on Kezar as his head poked through the curtain while the tiger cubs ran all over stage. At the end

of the show, I had to chase one cub from under Johnny's desk. On the way home, all the cubs and Kezar slept, having enjoyed themselves immensely.

Dog Meets Cat: When you introduce your cat to a new dog, your cat may become distant, aggressive, and stop using his litter box. You should introduce the new dog using the same procedures as in "Cat Meets Cat," but there are some extra considerations:

- Make sure you give your cat quality time, without the dog.
- If your cat becomes aggressive, he may be feeling defensive about the dog's crowding his territory. Provide him with places that the dog can't go, especially if it's a favorite sunning spot or resting area.
- Make sure the litter box is off-limits to the dog.

Other Types of Pets and Cats: American writer Muriel Beadle says, "Young kittens assume that all other animals are cats, approach them with jaunty friendliness and invite them to play." Cats have been known to befriend many different animals—rabbits, birds, guinea pigs, gerbils, and even mice. They have also been known to go out of their way to ingest their new friends. You need to be careful of this when introducing your cat to a smaller pet. Cats may react differently to small animals since a little animal's movement—the hopping of a rabbit or the stop-and-go jerkiness of a mouse—can excite and entice them. If you want to introduce your cat to smaller pets, it's best to start when your cat is young.

- Start slowly and try to teach your cat to be gentle with the smaller animal by reinforcing her when she is gentle and saying a stern "no" when she's too rough.

- Once they seem relaxed around each other, you can allow them to interact freely. But keep this play supervised and make sure that your cat's attitude doesn't change when she grows up. A playful kitten can become a predator cat.

The tiger cubs have been mixed with a great variety of animals—orangutans, dogs, chimpanzees, and even goats and sheep. I remember a play session among four tiger cubs, a dog (Kezar), one goat, and a gibbon. They all played in their own special way: the goat running away kicking up his heels; the gibbon jumping down on top of the tigers from an overhanging branch, or hanging upside down and grabbing the tigers' legs and tails, then scurrying up the tree for security; the tiger cubs batting up at the gibbon with their oversized paws and chasing the goat; and Kezar being satisfied lying there watching all the activity and letting each animal (except the goat) playfully jump, tug, and roll on him.

We introduce the tigers to other animals in order to stimulate them and socialize them to as many different situations as possible. This keeps them from being bored and helps round out their temperaments when reacting to new circumstances in the future. We don't, however, continue to mix these animals when the tigers are older as the tigers would become too rough.

Cats are usually gentle creatures and can get along with smaller animals. However, you must condition them to control their natural predatory instincts. And you must constantly supervise them when they're playing with smaller animals. Before you introduce your cat to a small animal, consider whether this will be more than a onetime occurrence; whether it will satisfy your cat's curiosity and stop her from trying to get at your small pet again; and whether it will help or hinder the harmony around the house. The last thing you want to do is activate the predator response in your cat.

Are All Cats Able to Mix with Other Pets?

Some cats' temperaments make it extremely hard for them to mix with other cats or animals, as in the case of Kimra, a very striking-looking tiger. At a young age she was mixed with another tiger, Kehei. We controlled the situation so they could get used to each other while not hurting each other. At first Kehei would get upset when Kimra came over—he was younger and lacked confidence, and so was overprotective. In time he calmed down, but as Kimra got older she became more aggressive and would take advantage of Kehei. She would constantly sneak up on him; what would start as playful behavior soon turned into trouble. As she got more aggressive, we had to separate them. We tried putting her with Serang, an easygoing male. But her antics were set. She continued to sneak up on her companions, and that only meant torment to poor Serang! Even though Serang outweighed her by more than two hundred pounds, that didn't stop Kimra.

Kimra just doesn't get along with any of the other tigers and is hostile toward them even if they grossly outweigh her. She now prefers human contact but at times even with us can have a "take it or leave it" attitude. Kimra is now content

to kick up her heels in green grassy areas with me and the other trainers; we are there to interact with her—if she so pleases.

We need to accept our pet cats for what they are. If your cat has a disposition similar to Kimra's, you need to respect her for the way she is and not force her to mix with another cat or pet. But if you are consistent with your conditioning methods, all but the most difficult of felines can learn to *tolerate* one another's presence—even though they may choose to be in a different room whenever possible.

COMPANY COMING AND NEW HOUSEHOLD MEMBERS

When people come over to your house, does your cat take off and hide? Then when everyone is gone, does he come back to your side? The cat who disappears when company comes often didn't have much exposure at an early age to people or hasn't experienced many different situations. It's easy to think that since your cat is so friendly with you he will be that way with everyone. That's not necessarily the case, even with cats who when young have socialized with many different people. To make your cat more congenial around company, try the following:

- Sit near your guest, but not so close as to scare your cat away.
- Call your cat and reinforce him with his favorite treat. If he hesitates to come over, move farther away from your guest.
- Once your cat is relaxed, slowly work your way closer to your guest, and have your guest offer a food reward to your brave cat.

- Practice this procedure consistently, giving your cat ample praise and attention.

The more your cat is around other people, the more he will feel comfortable and will be conditioned to them. If you don't want your cat to run every time someone comes into the house, you need to socialize him and expose him to people frequently. Just like any behavior, socializing needs to be *maintained*.

New Roommate or Spouse in the House: The addition of a new household member, whether it's a roommate, a spouse, a relative, or a short-term guest, is a change in routine for your cat. Your cat, as always, becomes alert when someone new comes into her territory. If she has been exposed to many people in the past, this change will be much easier. If your cat is not used to being around people, then she will look to you as the mother figure for protection. You need to be very supportive and give her ample attention, being careful not to make her jealous of this new member of your household, especially if it's a spouse.

When working with tigers, every trainer needs to be consistent with commands, playtime, quiet time, rewards, and feeding. When we introduce new people to the tigers, we first show and teach them everything having to do with the tigers. It will be helpful if you inform the new household member as much as possible about your feline companion, including

- Feeding time
- Regular quiet time and playtime
- Favorite hiding spots

- Preferred spots to be scratched
- Signals and calls you use
- Veterinarian's name and phone number

The more informed the new person is about your cat, the sooner they will come to know each other and be able to develop a relationship. The person will need to start out slowly and be aware of being perceived as a threat to your cat's territory or as competition for your attention. In the beginning you can do the same with the new person as you would making your cat comfortable with a guest, with the new person offering food rewards. Another thing is for that person to share quiet time with your cat.

New Baby—Cats and Infants: We know that bringing home a new baby changes *everyone's* routine. And it's very easy for a cat to be forgotten when a new baby first arrives. This is an extreme change for your cat and a time when you could see a tremendous change in his behavior. Cats are usually very gentle animals and need only some guidance when around a newborn baby. Look for any signals your cat is giving you signifying that he is upset:

- Destructive or aggressive behavior means "I'm bored from lack of activity and I need playtime."
- Not using the litter box or spraying means "I lack attention and have had enough break in my routine, how about just some quiet time together?"
- Being an annoyance around the baby means "I'm jealous and figure if I hang around this squirming thing, I'll get noticed!"

These signals are simply telling you your cat could use some tender loving care himself. Even in the most demanding times, try to take some time to include him in your daily activity with your baby and let him know he is important in your life. Cats and babies can develop wonderful relationships.

Cats and Children: Children can be rough on cats. It's up to you to set rules governing your child's behavior around your cat. Consistency is a major problem with children when it concerns your cat's behavior. Set "cat rules" for your kids to follow, which will help you and the cat:

Rule 1. Tell them *never* to hit the cat for any reason, even if he bites them. You need to teach your children to be gentle with the cat and show them how to pet him softly with an open hand.

Rule 2. No squeezing, no pulling the tail, no fast movements, no teasing, no squealing or shouting, no grabbing.

Rule 3. The cat is off-limits when he is sleeping, eating, or using the litter box.

These simple rules will help you all live peacefully under one roof. And what about tigers and children? Hmmmm, children are small, just about the perfect size for tiger prey. They move quickly and are erratic, just like a target the tiger would stalk in the wild. They can scream loudly, just like untamed game would. Do we think we should introduce children to tigers? Naaw.

cat conservation: we can make a difference

*It is impossible for a lover of cats to banish these
alert, gentle, and discriminating little friends, who
give us just enough of their regard and complaisance
to make us hunger for more.*
Agnes Repplier

CAT CONSERVATION

With hundreds of thousands of unwanted domestic cats abandoned
every year and only five of the eight subspecies of tigers still existing
today, is there anything that we can do to help? *Yes!* And we all need
to take *action* right away to do something for the welfare of the domestic cat and the preservation of the cat in the wild.

More cats die from behavior problems than from feline
leukemia. That's because a misbehaving cat is often given up by its
owners to an animal shelter. Because of the surplus of cats, that
usually means death. *Millions* of cats are euthanized every year in
animal shelters. And many wound up there because their owners did

not spend the necessary time with them to condition their behavior to ensure a happy and natural relationship between owner and cat.

We can all do our part to decrease these needless deaths by taking the responsibility of owning a pet very seriously. Be completely honest with yourself on your time limitations. The harsh truth of the matter is that if you do not have quality time to spend with your feline, you may be contributing to his death. No one likes to think of such a horrible outcome for his pet, but it happens to millions of cats a year.

STARTING IN YOUR OWN BACKYARD—SPAYING AND NEUTERING

We can easily contribute to the overpopulation solution by making sure that our cats are fixed as soon as they are old enough. The only reason not to have your pet fixed is if you are a professional breeder of cats. The following shocking information is provided by the Humane Society of the United States:

Save a life by adopting an animal from your local animal shelter!

"Cats have become too much of a good thing. The growing popularity of cats as companions, and the fact that cats are prolific breeders, have combined to send the cat population soaring in the last decade. The Humane Society of the United States estimates that cats now comprise more than half of the animals turned in to shelters nationwide. Based on estimated figures, it's safe to say that means *at least* four million cats. And a staggering 75% of the adoptable cats in shelters are euthanatized because there just aren't enough good homes. But this is one tragic problem that can be solved if every pet owner takes the simple step of spaying or neutering their cats. Every single litter adds directly to the problem of pet overpopulation. In fact, by mathematical progression, if a pair of cats has two litters per year, their offspring over seven years could account for 420,715 cats! And there are other good reasons to spay and neuter. Neutered male cats are less inclined to spray or roam. Unspayed female cats in heat can cry incessantly and attract unneutered male cats. Spaying and neutering reduce or eliminate such health problems as uterine, ovarian, breast and prostate cancer. Some people cite the cost of spaying or neutering as a deterrent. But when you add up the costs of *not* spaying and neutering in terms of the lives and deaths of innocent animals, the dollar amount is a small price to pay.'' The Humane Society of the United States is a leader in the fight against pet overpopulation.

TWO UNCONTROLLED BREEDING CATS—2 LITTERS PER YEAR, 2.8 SURVIVING KITTENS PER LITTER—WITH A 10-YEAR BREEDING LIFE, IN 10 YEARS MULTIPLY TO 80,399,780 CATS.
●

Every day in the United States, more than seventy thousand puppies and cats are born. When this number is compared to the ten thousand human births each day, it's clear that there *never* could be enough homes for all of these pets. But don't look at it as just a problem of numbers—every single pet is an individual life.

Here are some more advantages of spaying and neutering:
- Spayed and neutered dogs and cats live longer, healthier lives.
- Spayed and neutered pets are better, more affectionate companions.
- Spayed and neutered animals are less likely to bite. Unaltered

animals often exhibit more behavior and temperament problems than do those that have been spayed or neutered.

• Neutered males are less likely to run away or get into fights.

There are many organizations and people working hard to encourage spaying and neutering and pet protection methods. Find the ones in your area and contribute your time and money. For more suggestions, refer to the Appendix.

ENVIRONMENTALLY SAFE PET PRODUCTS AND PRACTICES

With the increasing number of cats kept as pets, there is also an increase in the impact cat products have on our environment. Environmental issues are no longer a fashionable issue but a serious reality about which each of us needs to be concerned. It is time that we all contribute to help our planet's situation.

There are many environmentally friendly pet products and practices that are simple to use and implement. One of the worst and most wasteful pet products is cat litter containing clay. The open mining of clay depletes a natural resource and destroys the environment—and clay is not biodegradable and must be disposed in landfills. There are many cat litters that do not contain clay that work just as well and are even more convenient since you can flush them down the toilet! Check the label and buy only clayless brands. Your cat won't mind a bit as long as you change to litter that is not scented at first. Some cats can be finicky about different scents in their litter box, especially in the beginning. Soon you can switch to a clayless brand that has natural deodorizers and perfume if you wish.

Purchasing individual-serving packages of cat food is extremely wasteful. If you do need to buy individual servings for your cat, make sure you buy them in aluminum cans and always recycle the cans. Otherwise buy your cat food and kitty litter in bulk. There are many brands that use reusable and recyclable containers. Discarded packaging contributes to one third of the earth's waste. Just think what a difference it would make in cutting down the waste in overcrowded landfills if every pet owner was responsible when purchasing pet food.

Another way we can help our environment is by refusing to purchase harsh chemicals as cleaning agents. When you clean and disinfect a litter box, you can use these harmless solutions:

- One half cup of borax (a natural mineral that kills mold and bacteria) dissolved in one gallon of hot water, or
- Two cups of boiling water poured over two cups of fresh thyme, an antiseptic herb, and steeped for ten minutes, then strained and cooled. (Store this in plastic bottles.)

These two solutions are also good disinfectants for all your bathroom and kitchen surfaces.

Remember, your cat spends a lot more time on the floor and on surfaces in your house that you do. Harsh cleaning chemicals can be toxic to him, so here are some household cleaners you can use that won't hurt your cat and are good for the planet:

- **All-purpose Floor Cleaner**—one cup of vinegar in a pail of water. For greasy floors, add one fourth cup washing soda and one tablespoon of vegetable-oil-based soap to every two gallons of hot water.
- **Floor Polish** (for linoleum floors)—one part baking soda to three parts water.

- **Carpet Cleaner and Deodorizer**—two parts cornmeal (which absorbs grease) mixed with one part borax. Sprinkle generously over a rug and let sit for an hour. Then vacuum.
- **Window and Mirror Cleaner**—one part vinegar mixed with four parts water. Or combine one cup of cold strong black tea with three tablespoons of vinegar. Store these solutions in spray bottles.
- **Furniture Water Spots Remover**—a little white toothpaste massaged on the spot, let dry, and buffed with a clean cloth.
- **Wood Surface Scratches or Stains Repair**—the "meat" of a walnut rubbed into the scratch or stain.

These products are inexpensive and easy for you to mix yourself and use on a daily basis. There are also many products you can purchase that are environmentally friendly. When you buy any shampoos, flea collars (there are herbal ones available), cat food, catnip (organic of course), insecticides, and toys, look for brands that are environmentally safe. Be careful when purchasing tuna cat food that the tuna does not come at the expense of a dolphin's life. Many brands of "tuna-flavored" cat foods still contain tuna that is *netted*. Netting contributes to the deaths of hundreds of thousands of dolphins each year. Two brands that carry "dolphin safe" varieties are Fancy Feast and Alpo.

If your cat is an indoor/outdoor cat it is also very important to stay informed about developments in your neighborhood. Be on the watch for any pest control programs that involve poisons. Also keep informed about any factories that dispose of hazardous chemicals. As an activist in your community, you can make a difference for the health of your cat, your family, and the planet.

THE ENDANGERMENT OF THE TIGER

A large crowd is watching as Mohan comes walking over. He hops up and puts his big paw on my shoulder. This allows him to stretch out his five-hundred-pound frame to its full ten feet. His reward is a small carton of milk, which he laps up with his oversized tongue. With Mohan's help I now have everyone's attention and a perfect opportunity to talk about tigers in the wild and the urgent need to protect them.

People are always amazed at the tigers' physical size and power. Some find it hard to believe that they not only love to swim but will seek out water. But the strongest reaction, as always, comes when people hear that three of the eight tiger subspecies are now thought to be extinct—the Caspian, the Javan, and the Balinese. Tigers used to range across much of Asia from Siberia through China, down into Indochina, the Malay Peninsula, and the Indonesian islands, across India and Nepal. These are the recent pitiful facts:

- The last tiger in Bali was seen in 1940.
- The last glimpse of three tigers in Java was in 1980.
- There have been no sightings or evidence that the Caspian tiger has survived, since 1972.
- The Chinese tiger population is now down to between thirty and fifty cats, in such scattered pockets, and under such furious habitat pressures, that the future does not look good for this beautiful animal.

IT WAS ONCE THOUGHT THAT THE NUMBER OF TIGERS SPREAD ALL THROUGH ASIA EXCEEDED 100,000. NOW THERE ARE ONLY BETWEEN 6,000 AND 9,000 IN 12 COUNTRIES IN ASIA. OF THE 5 SUBSPECIES THAT REMAIN, THE BENGAL TIGER IS THE MOST ABUNDANT.

●

This is the latest status of the tigers in Asia.

WORLD TIGER POPULATION		
Type of Tiger	Minimum Number	Maximum Number
Bengal	4,450	5,950
Siberian	230	400
Chinese	30	50
Sumatran	500	1,000
Indo-Chinese	1,160	1,520
Total	6,370	8,920

KOREA IS STILL CALLED "LAND OF THE WHITE TIGER," BUT IT IS DOUBTFUL THAT ANY TIGERS SURVIVE THERE TO DATE.

•

Mohan and the others at Tiger Island are all Bengal tigers, which are the most abundant of all the remaining tigers, including those in the wild and in captivity. The reason for this is that when, in 1969, India became aware of the drastic reduction in the tiger population, and discovered that only about two thousand Bengal tigers remained, it became determined to take action.

In 1973, Project Tiger was launched with the Indian government and the World Wildlife Fund, establishing eight tiger reserves. Each reserve was originated with the intent that not only would the tiger be protected, but every plant and animal there also. Strong strategies from major conservation bodies protected the tigers from the constant pressure of development that would encroach on these preserves. The decline of the tiger is a direct result of man's increasing demand for land and natural resources. The tigers' natural environment is slowly being cleared to make way for mankind. And with each step, we bring the tiger closer to the brink of extinction. Now there are seventeen prime tiger reserves in India totaling some twenty-five thousand square kilometers of protected lands. Setting aside biological reserves tends to conserve not only the tigers but entire wild ecosystems.

Due to those nineteen years of hard work and commitment, the tiger was brought back from the brink of extinction. The wildlife conservation programs have been so successful that now the parks and sanctuaries are no longer big enough. The "buffer zones" that before kept distance and safety between people and tigers are now areas needed to provide wood and farming. Because of the growing demand of land for both man and tiger, deadly encounters have occurred. Fifty to sixty people are killed annually by tiger attacks in India. Man and tiger cannot share the same range of area. To ensure the maintenance of a viable population of tigers, the area must be off-limits to people.

Because of the combination of all of these efforts, hope for the tiger in India has increased. A census in 1979 indicated the tiger population had risen to three thousand and, ten years later, in 1989, had grown to over four thousand tigers. The increase of the number of special reserves for tigers to seventeen clearly shows that conservation efforts are succeeding, but there is still a long way to go. Facilities like Tiger Island also play an important role in the preservation of these beautiful large cats.

THE SURVIVAL OF EXOTIC CATS
IN AMERICA

Asia isn't the only continent with exotic cats at the brink of extinction. Right here in North America, the Florida panther is an endangered species. This large cat with many names—puma, mountain
lion, cougar, panther, painter, and catamount—ranges more widely
than any other terrestrial mammal in the Americas. But the Florida
panther is on the brink of extinction.

The Florida panther is one of seven subspecies of exotic cats in North America. Recent population studies show that in 1990 there were only between thirty and fifty of these beautiful cats left. At one time this large cat's area ranged from Arkansas to South Carolina to Florida. Its territory now consists of five thousand square miles centered in the Everglades National Wildlife Refuge.

Because of man's constant encroachment, the panther's range and population has been reduced so much that inbreeding has become the norm. It is so severe that the panther is now showing several abnormal traits.

Another problem is the hybridization (mixing) of the Florida panther with the Central or South American panther. Because of this, the Florida panther could be taken off the endangered species list—the list applies only to recognized species and subspecies.

The only solution is a breeding program for panthers in captivity which could be the panther's last chance at survival in the wild. The goal is 130 breeding animals in a combination of wild and captive environments by the year 2000 and 500 by 2010. This would assure a 95 percent possibility that the cat will survive in the wild for one hundred years.

If you want more information or would like to help, write to the Florida Panther Trust Fund, Florida Game and Freshwater Fish Commission, 620 Meridian Street, Tallahassee, Florida 32399–1600.

A TENDER TIME WITH A TIGER

One of the most memorable moments I've shared with the tigers involved a young girl. Her mother stopped me one day when I was walking Kehei, who is completely blind, on a leash. She asked if her little girl might be able to pet Kehei; she explained that her daughter, like Kehei, also has a disability and had been blind since birth. All of her life the girl has had to have everything, even the appearance of a tiger, described to her. As I watched this young girl's fingers run through Kehei's fur, I realized she was using all of her senses to feel, hear, smell, and understand the power and beauty of this tiger. I think she has now a clearer picture of a tiger than almost anyone I know.

We made a commitment to Kehei and spent quality time and hours of training with him, so he trusted us and we were able to build a relationship with him. Because of this, he stayed relaxed while this young girl stroked him, giving her the opportunity to have the true image of a tiger, an image that will stay with her for the rest of her life. I know real men aren't supposed to cry (especially those guys that spend their day with those ferocious tigers), but seeing the look of appreciation on that little girl's face, I did.

If each of us gains a small amount of this young girl's appreciation for the feline, by making a commitment to spend quality time with our cats at home and doing whatever we can to help protect stray cats and cats in the wild, we will enrich our own lives immeasurably.

WHAT TO DO?

Make a commitment to become well informed and to contact conservation groups and local animal shelters. There are also many international organizations dedicated to the protection of exotic cats and their habitat in the wild. If you are interested and want to know what you can do to help, contact the organizations listed in the Appendix and ask them for the names and addresses of the many other wonderful organizations in your area and around the world.

If you have any stories you would like to share with us, please write to Tiger on Your Couch, P.O. Box 14, Breckenridge, Colorado 80424.

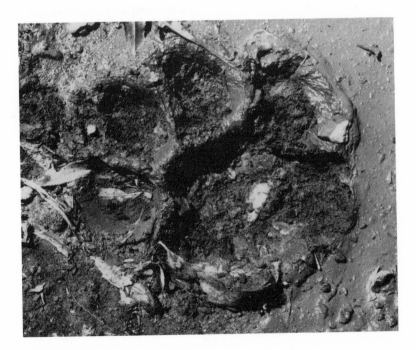

appendix

National Groups

American Wildlands
7500 E. Arapahoe, Suite 355
Englewood, CO 80112
(303) 771–0380

Defenders of Wildlife
12444 Nineteenth Street NW
Washington, DC 20036
(202) 659–9510

Florida Panther Coordinator
117 Newins-Ziegler Hall
University of Florida
Gainesville, FL 32611
(904) 392–1861

Florida Panther Research and
  Management Trust Fund
Florida Game and Fresh Water
  Fish Commission
620 S. Meridian Street
Tallahassee, FL 32301
(904) 488–3831

Mountain Lion Preservation
  Foundation
P.O. Box 1896
Sacramento, CA 95809
(707) 442–2666

National Audubon Society
950 Third Avenue
New York, NY 10022
(212) 832–3200

Wilderness Society
900 Seventeenth Street NW
Washington, DC 20006–2596
(202) 833–2300

International Groups

African Wildlife Foundation
1717 Massachusetts Avenue NW
Washington, DC 20036
(202) 265–8393

Canadian Wildlife Federation
1673 Carling Avenue
Ottawa, Ontario
Canada K2A 3Z1
(613) 725–2191

International Society for
  Endangered Cats, Inc.
4638 Winterset Drive
Columbus, OH 43220
(614) 451–4460

National Geographic Society
17 and M Street NW
Washington, DC 20036
(202) 857–7000

National Wildlife Federation
1400 Sixteenth Street NW
Washington, DC 20036–2266
(202) 637–3700

Rainforest Alliance
270 Lafayette Street, Suite 512
New York, NY 10012
(212) 941–1900

Sierra Club
730 Polk Street
San Francisco, CA 94109
(415) 776–2211

World Wildlife Fund/The
  Conservation Foundation
1250 Twenty-fourth Street NW
Washington, DC 20037
(202) 293–4800

Animal Welfare Organizations

American Society for the Prevention
  of Cruelty to Animals
441 E. Ninety-second Street
New York, NY 10128
(212) 831–6006

The American Humane Association
63 Inverness Drive East
Englewood, CO 80112
(303) 792–9900

Cornell Feline Health Center
New York State College of
 Veterinary Medicine
Cornell University
Ithaca, NY 14853
(607) 253–3414

The Delta Society:
Interactions of People, Animals
 and the Environment
P.O. Box 1080
Renton, WA 98057
(206) 226–7357

Friends of Animals Inc.
P.O. Box 1244
Norwalk, CT 06856
(203) 866–5223

The Humane Society of the
 United States
2100 L Street NW
Washington, DC 20037
(800) 372–0800

Morris Animal Foundation
45 Inverness Drive East
Englewood, CO 80112
(707) 790–2345

National Humane Education Society
Elisabeth Guillet Vlk
15B Catoctin Circle SE, #207
Leesburg, VA 22075
(804) 777–8319

Progressive Animal Welfare Society
15305 Forty-fourth Avenue West
Lynnwood, WA 98046
(206) 743–3845

Robert H. Winn Foundation for
 Cat Research
1309 Allaire Avenue
Ocean, NJ 07712
(908) 528–9797

Tree House Animal Foundation Inc.
1212 W. Carmen Avenue
Chicago, IL 60640
(312) 784–5488

Shelters Meeting American Humane Association Standards of Excellence

Because not all animal shelters, humane societies, and SPCAs are related across the nation, there are many standards. The American Humane Association has a Standard of Excellence program that recognizes shelters for exceeding American Humane Association minimum standards in five categories: facilities, humane euthanasia, community relations and education, and planning. If a shelter is not listed for your area, contact a vet for a strong recommendation.

Arizona

Arizona Humane Society
P.O. Box 47240
Phoenix, AZ 85068
(602) 997–7585

The Pima Animal Control Center
4000 N. Silverbell Road
Tucson, AZ 85745
(602) 743–7550

Humane Society of the White
 Mountains
P.O. Box 1070
Pinetop, AZ 85935
(602) 368–5295

Arkansas

Good Shepherd Humane Society
P.O. Box 285
Eureka Springs, AR 72632
(501) 253–9188

California

San Diego Humane Society
887 Sherman Street
San Diego, CA 92110
(619) 299–7012

Colorado

Boulder County Humane Society
2323 Fifty-fifth Street
Boulder, CO 80301
(303) 442–4030

La Plata County Humane Society
P.O. Box 2164
Durango, CO 81302
(303) 259–2847

Cat Care Society
5985 W. Eleventh Avenue
Lakewood, CO 80214
(303) 239–9690

Denver Dumb Friends League
2080 South Quebec Street
Denver, CO 80231
(303) 671–5212

Jefferson Animal Shelter
4105 Youngfield Service Road
Golden, CO 80401
(303) 278–7575

Longmont Humane Society
9595 Nelson Road, Box G
Longmont, CO 80501
(303) 772–1232

Florida

Halifax Humane Society
P.O. Box 624
Daytona Beach, FL 32117
(904) 274–4710

Humane Society of St. Lucie County
P.O. Box 3661–Savannah Road
Ft. Pierce, FL 34982
(407) 461–0687

Humane Society of Broward County
2070 Griffin Road
Ft. Lauderdale, FL 33312
(305) 989–3977

Alachua County Animal Control
3400 NE. Fifty-third Avenue
Gainesville, FL 32609
(904) 336–2333

Port St. Lucie Animal Control
450 SW. Thornhill Drive
Port St. Lucie, FL 34980
(407) 871–5042

A.R.L. of the Palm Beaches
3200 North Military Trail
West Palm Beach, FL 33409
(407) 686–3663

Georgia

Atlanta Humane Society
981 Howell Mill Road
Atlanta, GA 30318
(404) 875–5331

Hawaii

Hawaii Island Humane Society
P.O. Box 939
Keaau, HI 96749
(808) 966–8161

Maui Humane Society
P.O. Box 397
Kihei, Maui, HI 96753
(808) 877–3680

Illinois

Anti-Cruelty Society
157 W. Grand
Chicago, IL 60610
(312) 644–8338

The Wayne County Humane Society
P.O. Box 254
Fairfield, IL 62837
(618) 847–2981

Anderson Animal Shelter
100 S. Lafox
South Elgin, IL 60177
(708) 697–2882

Indiana

Bartholomew County Humane
  Society
P.O. Box 1088
Columbus, IN 47201
(812) 372–6063

Humane Society of St. Joseph
  County
2506 Liberty Drive
Mishawaka, IN 46545
(219) 255–4726

Iowa

Animal Rescue League of Iowa
5452 NE. Twenty-second Street
Des Moines, IA 50313
(515) 262–9503

Maryland

Montgomery County Animal Control
14645 Rothgeb Drive
Rockville, MD 20850
(301) 279–1066

Michigan

Humane Society of Genesee County
3075 Joyce Street
Burton, MI 48529
(313) 744–0511

Minnesota

St. Croix Animal Shelter
13342 Fortieth Street S
Afton, MN 55001
(612) 436–7366

Humane Society of Ramsey County
1115 Beulah Lane
St. Paul, MN 55108
(612) 645–7387

Animal Humane Society
845 Meadow Lane North
Golden Valley, MN 55422
(612) 522–4325

Missouri

Central Missouri Humane Society
616 Big Bear Boulevard
Columbia, MO 65202
(314) 443–3893

Humane Society of Missouri
1210 Macklind Avenue
St. Louis, MO 63110
(314) 647–8800

Montana

Missoula County Humane Society
1105 Clark Fork Drive
Missoula, MT 59802
(406) 549–3934

North Carolina

Animal Protection Society of
  Orange County
1081 Airport Road
Chapel Hill, NC 27514
(919) 967–7383

Nebraska

Animal Control/ City of Lincoln
2200 St. Marys Avenue
Lincoln, NE 68502
(402) 471–7900

New Jersey

St. Hubert's Giralda
P.O. Box 159
Madison, NJ 07940
(201) 377–2295

New York

Town of Hempstead Animal Shelter
3320 Beltagh Avenue
Wantagh, NY 11793
(516) 785–5220

Ohio

Hamilton County SPCA
3049 Colerain Avenue
Cincinnati, OH 45223
(513) 541–6100

Toledo Humane Society
1320 Indianwood Circle
Maumee, OH 43537
(419) 891–0705

Oregon

Multnomah County Animal Control
24450 W. Columbia Highway
Troutdale, OR 97060
(503) 667–7387

Pennsylvania

Womens Humane Society of
  Pennsylvania
3025 W. Clearfield
Philadelphia, PA 19132
(215) 225–4500

ARL of Western Pennsylvania
6620 Hamilton Avenue
Pittsburgh, PA 15206
(412) 661–6452

Rhode Island

Robert Potter League for Animals
P.O. Box 412
Newport, RI 02840
(401) 846–8276

South Carolina

Aiken SPCA
401 Wire Road
Aiken, SC 29801
(803) 648–6863

Tennessee

Washington County/Johnson City AC
525 Sellis Avenue
Johnson City, TN 37604
(615) 926–8769

Texas

Humane Society of Austin &
  Travis County
P.O. Box 1386
Austin, TX 78767
(512) 478–9325

Brazos Animal Shelter
P.O. Box 4191
Bryan, TX 77805
(409) 775–5755

SPCA of Texas
362 S. Industrial Boulevard
Dallas, TX 75207
(214) 651–9611

Houston SPCA
519 Studmont
Houston, TX 77007
(713) 869–8227

Waco Humane Society
2032 Circle Road
Waco, TX 76706
(817) 754–1454

Utah

Humane Society of Utah
4613 S. 4000 West
Salt Lake City, UT 84120
(801) 968–3548

Virginia

Fairfax County Animal Control
4500 West Ox Road
Fairfax, VA 22030
(703) 830–3680

Richmond SPCA
1800 Chamberlayne Avenue
Richmond, VA 23222
(804) 643–6785

Virginia Beach SPCA
3040 Holland Road
Virginia Beach, VA 23456

Washington

Humane Society Seattle—King
  County
13212 SE. Eastgate Way
Bellevue, WA 98005
(206) 641–0080

Benton-Franklin Humane Society
P.O. Box 2532
Pasco, WA 99301
(509) 545–9301

Kitsap Humane Society
9167 Dickey Road NW
Silverdale, WA 98383
(206) 692–6977

The Humane Society for
  Tacoma & Pierce County
2608 Center Street
Tacoma, WA 98409
(206) 383–2733

Wisconsin

Fox Valley Humane Association
3401 W. Brewster Street
Appleton, WI 54914
(414) 733–1717

Wyoming

Metro Animal Control & Welfare
200 N. David
Casper, WY 82601
(307) 235–8398

permissions

Grateful acknowledgment is made for permission to reprint the following:

From *The Cat Diary*, edited by the Running Press Staff, copyright © 1990 by Running Press, courtesy of Running Press Book Publishers.

Page 15, Jean Burden; page 91, Fernand Mery; page 93, François-René de Chateaubriand; page 113, Janet F. Faure; page 124, Wesley Bates; page 188, Muriel Beadle.

From *The Literary Cat*, copyright © 1990 by Running Press, courtesy of Running Press Book Publishers.

Page 29, Jean Cocteau; page 38, Agnes Repplier; page 44, Eric Gurney; page 48, Fernand Mery; page 51, Sir Compton Mackenzie; page 60, St. George Mivart; page 73, Monica Edwards; page 197, Agnes Repplier.

From *The Cat Notebook*, copyright © 1981 by Running Press, courtesy of Running Press Book Publishers.

Page 135, Arthur Weigall; page 165, F. A. Paradis de Moncrif; page 177, Jean Jacques Rousseau.

From *Cats Are Cats*, compiled by Nancy Larrick, copyright © 1988 by Philomel Books, courtesy of Eve Merriam.
Page 45, "The Stray Cat," by Eve Merriam.

From *Endangered Species Tigers* by Peter Jackson, copyright © 1990 by Quintet Publishing Ltd., courtesy of Quintet Publishing Ltd.
Page 204, "World Tiger Population."

From The Humane Society of the United States, information on cat overpopulation.

From the American Humane Association, the list of shelters that met the Standards of Excellence.

From *The Florida Panther* booklet by the Florida Power Company, copyright © 1988 by the Florida Power Company, courtesy of the Florida Power Company, brief excerpts.

From Jennifer Fox, owner of Yoga Fitness, 350 Molino Avenue, Mill Valley, CA 94941, (415) 381-YOGA, meditation with—and for—your cat and partner breathing with your cat.

From Dr. Alfred Kissileff, D.V.M., information on flea control.

From Gillian Rice, Creator of Pussy Cat Pool (1-800-9555-CAT), for photographs.

photograph credits

Page 14: Darryl Bush. **Page 15:** "Esi," Patrick Martin-Vegue/Marine World Africa USA. **Page 17:** *left*, "Mohan," Patrick Martin-Vegue/Marine World Africa USA; *right*, Darryl Bush. **Page 19:** "Mohan," Gregg Lee/Marine World Africa USA. **Page 20:** *top*, Gillian Rice; *bottom*, "Baghdad," Patrick Martin-Vegue/Marine World Africa USA. **Page 23:** "Uki and Hobbs," Deb Reingold. **Page 24:** *left*, "Jai and Sampson," Patrick Martin-Vegue/Marine World Africa USA; *right*, "Rakon and Kali," Patrick Martin-Vegue/Marine World Africa USA. **Page 25:** *left*, "Gregg Lee, Bill Fleming, and Patrick Martin-Vegue with Jai, Tara, and Sampson," Darryl Bush/Marine World Africa USA; *right*, "Kismet (adult) and Liza," Patrick Martin-Vegue/Marine World Africa USA. **Page 27:** "Uki," Deb Reingold. **Page 28:** "Céilidh," Caroline Sykes. **Page 30:** "Hobbs," Deb Reingold. **Page 34:** "Uki and Hobbs," Deb Reingold. **Page 35:** "Rakon and Kali," Gregg Lee/Marine World Africa USA. **Page 37:** *top*, "Tiger in a basket: Zeck, Yuhasz, Gallagher, Wilde, and Toklas," Phillip Blackmon; *bottom*, "Sampson (white) and other cubs," Courtesy Marine World Africa USA. **Page 38:** Gillian Rice. **Page 41:** "Esi," Patrick Martin-Vegue/Marine World Africa USA. **Page 44:** "Uki," Deb Reingold. **Page 45:** "Uki and Hobbs," Deb Reingold. **Page 47:** Darryl Bush. **Page 50:** "Kismet and cubs," Patrick Martin-Vegue/Marine World Africa USA. **Page 53:** "Sappho (mother) and Zeck, Yuhasz, Gallagher, Wilde, and Toklas," Phillip Blackmon. **Page 59:** "Sampson and Jai," Darryl Bush/Marine World Africa USA. **Page 65:** "Rakon," Patrick Martin-Vegue/Marine World Africa USA. **Page 72:** "Bill Fleming and Tara," Darryl Bush/Marine World Africa USA. **Page**

74: "Uki and Hobbs," Deb Reingold. **Page 75:** "Uki," Deb Reingold. **Page 76:** *left*, Darryl Bush; *right*, "Kimra," Gregg Lee/Marine World Africa USA. **Page 79:** Darryl Bush. **Page 81:** "Hobbs," Deb Reingold. **Page 87:** "Sampson," Patrick Martin-Vegue/Marine World Africa USA. **Page 90:** "Uki and Hobbs," Deb Reingold. **Page 99:** "Timber," Darryl Bush. **Page 100:** "Tara and Jai," Courtesy Marine World Africa USA. **Page 104:** *left*, "Cinnie grooming," Darryl Bush; *right*, "Sampson, Jai, and Tara," Patrick Martin-Vegue/Marine World Africa USA. **Page 109:** "Uki," Deb Reingold. **Page 111:** "Serang," Darryl Bush/Marine World Africa USA. **Page 112:** "Meghan Nix, age six, and Tigger," Darryl Bush. **Page 119:** "Lectra," Darryl Bush. **Page 120:** "The tiger on your toilet. Worley demonstrates a no-no," Phillip Blackmon. **Page 125:** *left*, "Uki," Deb Reingold; *right*, "Tiger cub," Bill Fleming/Marine World Africa USA. **Page 128:** *left*, "Sampson with fish," Gregg Lee/Marine World Africa USA; *right*, "Tigger sipping water," Darryl Bush. **Page 131:** "Andy and Rakon," Darryl Bush/Marine World Africa USA. **Page 134:** "Diane Nix and Tigger," Darryl Bush. **Page 140:** Gillian Rice. **Page 147:** Darryl Bush. **Page 151:** "Kurby and Jake," Patrick Martin-Vegue/Marine World Africa USA. **Page 155:** "Mohan and Baghdad," Patrick Martin-Vegue/Marine World Africa USA. **Page 158:** *left*, "Uki and Hobbs," Deb Reingold; *right*, "Rakon," Courtesy Marine World Africa USA. **Page 162:** "Sampson and Tara," Patrick Martin-Vegue/Marine World Africa USA. **Page 164:** "Lestat maintaining his lonely vigil waiting to get in the car (this was the first time he'd seen snow and he didn't know what to make of it)," Phillip Blackmon. **Page 166:** "Lestat on top of a cage and Crawford inside. Traveling sales cats waiting for their human on a cold Ohio morning," Phillip Blackmon. **Page 172:** "Len Meyers with Tara," Patrick Martin-Vegue/Marine World Africa USA. **Page 176:** "Eight-ball with a dog," Gillian Rice. **Page 181:** "Kezar and tiger cubs," Courtesy Marine World Africa USA. **Page 186:** "Worley at eight weeks," Phillip Blackmon. **Page 193:** "Uki," Deb Reingold. **Page 195:** "Meghan and Tigger at a day care center," Darryl Bush. **Page 196:** "Kismet (adult) and Liza," Patrick Martin-Vegue/Marine World Africa USA. **Page 203:** "Sampson," Andy Goldfarb/Marine World Africa USA. **Page 206:** "Rakon, Serang, and Kali," Patrick Martin-Vegue/Marine World Africa USA. **Page 207:** Patrick Martin-Vegue/Marine World Africa USA.

color photograph credits

Page 1: Gillian Rice. **Page 2:** "Pokie," Caryl Lenahan. **Page 3:** "Mohan," Patrick Martin-Vegue/Marine World Africa USA. **Page 4:** "Rakon," Patrick Martin-Vegue/Marine World Africa USA. **Page 5:** "Worley surveying his territory," Phillip Blackmon. **Page 6:** "Uki and Hobbs," Deb Reingold. **Page 7:** "Uki," Deb Reingold. **Page 8:** "Tiger cub," Bill Fleming/Marine World Africa USA. **Page 9:** Darryl Bush. **Page 10:** "Serang," Patrick Martin-Vegue/Marine World Africa USA. **Page 11:** "Jai," Patrick Martin-Vegue/Marine World Africa USA. **Page 12:** Darryl Bush. **Page 13:** "Uki," Deb Reingold. **Page 14:** "Ollie with a tiger cub," Courtesy Marine World Africa USA. **Page 15:** "Bill playing with Rakon—careful conditioning keeps even the most rambunctious playtime safe and fun for tiger and trainer alike," Patrick Martin-Vegue/ Marine World Africa USA. **Page 16:** *top*, "Kismet," Patrick Martin-Vegue/Marine World Africa USA; *bottom*, "Baghdad," Patrick Martin-Vegue/Marine World Africa USA. **Page 17:** "Sampson, Tara, and Jai," Patrick Martin-Vegue/Marine World Africa USA.

abandoned cats, 29–30
abdominal area, 103
activity, 74, 80, 88, 146
    spraying and, 118
    see also playtime
adult cat, choosing as pet, 36–37
affection, from cat, 95, 98, 199
    conditioning for, 98–101
affection, giving to cat, 59
aggressive behavior, 69–71, 77–78, 88,
        93
air travel, 173
aloof behavior, 94, 95
aluminum foil balls, 85, 120
Animal Guinness Book of World Records,
        25
animal shelters, 29–30, 45, 197–199,
        209
Animal Welfare Institute, 143
antifreeze, 142, 150
appliances, 121
Asia, 18, 48, 203, 206
Auguste-René, François, 93

baby, new, 193
Baghdad, 12, 18, 25, 26, 70, 83,
        99–101, 102, 174
bags, paper, 85
bags, plastic, 120
Bali, 18, 203
Bates, Wesley, 124
baths, 18, 161–163
Beadle, Muriel, 188

behavior, 12, 180
    problems with, 12, 13, 197–198
    and spaying or neutering, 200
    unexpected, 69–71
    see also conditioning of cat's
        behavior; *specific behaviors*
Bengal tiger, 204
birds, 157, 188
bite marks, on cat, 103
biting, 34, 55, 69, 75, 77
    controlling of, 64–66
    predatory instinct and, 52
    and spaying or neutering, 199
bladder problems, 128, 130
bonding, 54, 94, 97–98
boredom, 79–84, 118
    stress and, 74, 88
breakable items, 33, 78–79, 124
breath, 39, 103
breathing, partner, 108–111
breeders, 45–46
Brown, Mary, 180
brushing, 32, 104–105
Burden, Jean, 15

cabinet doors, 132, 134–135
call, training cat to respond to,
        135–138
camouflage, 16–17
cancer, 199
car, traveling in, 170–172
car hood, 150
carpet cleaners, 120, 202

cat door, 147–148, 149
cat litter, 200, 201
catnip, 80, 83, 123, 202
change in routine, 129, 177–178
    see also new situations
Chateaubriand, François
        Auguste-René de
checkup, physical, 102–103
children, 194
Chinese tiger, 203
choosing your cat, 29–49
    adult cat, 36–37
    health check in, 39
    kitten, 37–38
    life-style and, 29–33
    personality in, 44
    two cats vs. one, 34–36
    where to find your cat, 45–48
chuff, 21–22, 96
claws, 64, 67–69, 103
    see also scratching
clay, cat litter made from, 200
cleaning products, 120, 201–202
climbing abilities, 22–23
clothes dryers, 120
coat, *see* fur
Cocteau, Jean, 29
cold weather, 150
collar, 143
color, as stimulus, 80
color and markings, 16–17
communication, 96–97
company, 191–194

conditioning of cat's behavior, 27,
    53–63, 114
  to accept travel box, 166–169
  affection in, 59
  to be cozier, 98–101
  best time for, with kitten, 37–38,
    56–58
  to come to dinner call, 151–153
  to come when called, 135–138
  consistency in, 58–59, 68, 88, 105
  to control biting, 64–66
  to keep claws in, 64, 68–69
  to keep off countertops, 133–134
  mother figure and, 53–54, 55, 58,
    68, 133, 135
  patience in, 60, 69
  playtime and, 73, 74, 75
  repetition in, 61
  rules for, 59–63
  to stop spraying, 118–119
  thinking like a cat and, 60–61, 122,
    124, 169
  time-outs in, 61
  time spent in, 60
  treats in, 132, 136, 138–139, 153
  unexpected behavior and, 69–71
  to use litter box, 115
conservation, 197–209
  environmentally safe products and
    practices in, 200–202
  of Florida panthers, 206–207
  spaying and neutering as part of,
    198–200
  of tigers, 203–205
  what to do for, 209
consistency, in conditioning, 58–59,
    68, 88, 105
countertops, jumping on, 132, 133–134
cozy cat, 98–101
cupboards, 132, 134–135

diet, see feeding, food
dinner call, 151–153
discipline, 62
  see also conditioning of cat's
    behavior
distemper, 175
dog, introducing cat to, 186–188
dolphins, 202

door, for indoor/outdoor cat, 147–148,
    149

ears, 39, 102
Edwards, Monica, 73
electrical cords, 120–121
energy, playtime and, 74, 78, 88, 89,
    114, 124
environmentally safe products and
    practices, 200–202
Esi, 40, 42–43
euthanasia, 12, 197–199
Everglades National Wildlife Refuge,
    207
excitement, 75–76, 77
exercise, 70, 78, 88, 142
  see also playtime
exotic cats, 206–207, 209
expenses, cat care, 31
eyes, 20–21, 39, 102

fan belts, 150
Faure, Janet F., 113
feeding, food, 126–135, 202
  advice on, 127–128
  dinner call and, 151–153
  finickiness and, 128, 129, 130
  hunting and, 157
  individual-serving packages of,
    201
  for indoor/outdoor cat, 145, 150
  kitchen and dinnertime tips for,
    130–135
  "nourishment no's" for, 130
  for outdoor cat, 150
  possessiveness and, 86
  schedule for, 128
  transitions in, 129
feline distemper, 175
feline leukemia, 175, 197
fights, cat, 103, 145, 153–155
  and spaying or neutering, 32, 153,
    154, 181, 200
finickiness, 128, 129, 130
fitness, 88–89
  see also exercise; playtime
fixing, see neutering; spaying
flea collars, 202
flea comb, 105–106

fleas, 103, 105–106, 161, 162
Flehmen response, 20
floor cleaner, 201
floor polish, 201
Florida panther, 206–207
Florida Panther Trust Fund, 207
foil balls, 85, 120
Fridge, 15, 83–84
fur, 17–18, 39, 102, 103
  shedding of, 32, 104
furniture, 78, 121–123
furniture polishes, 120
furniture water spots remover, 202
FUS (Feline Urological Syndrome), 130

games, see playtime; toys
garbage and trash, 121, 132, 150
garden, cat, 124–125
gerbils, 188
gifts, from cat, 156
greeting at the door, 137–138
grooming:
  brushing in, 32, 104–105
  self-, 18
guests in house, 191, 192
guinea pigs, 188
Gurney, Eric, 44

hair, see fur
hands, as toys, 75, 77
harness, 160
health:
  in choosing cat, 39
  and introducing cat to new pet, 180
  playtime and, 74, 88
  quiet time checkup for, 102–103
  and spaying or neutering, 199
hearing, sense of, 80, 82–83
heat, female cat in, 70, 129, 145, 154,
    199
hiding, in play, 82, 156
hormonal change, 70
household cleaners, 201–202
household hazards, 120–121
household members, new, 191–194
house plants, 124–126
housing, outdoor, 150
Humane Society of the United
    States, 198–199

hunting, 94, 145
  of birds, 157
hunting skills, 51, 52, 156
  playtime and, 75, 88, 146, 157

indoor cat, 113–139
  activity needed by, 74
  breakable items and, 33, 78–79, 124
  feeding of, see feeding, food
  gifts from, 156
  household hazards to, 120–121
  indoor/outdoor cat vs., 141–143
  plants and, 124–126
  scratching post for, 121–123
  training to respond to call, 135–138
indoor/outdoor cat, 113, 141–163
  bathing of, 161–163
  birds and, 157
  bite marks on, 103
  cat door for, 147–148, 149
  cold weather and, 150
  coming in and going out of house,
      147–149
  dinner call for, 151–153
  feeding of, 145, 150; see also
      feeding, food
  fights and, see fights, cat
  indoor cat vs., 141–143
  introducing to outdoors, 143–147
  litter box for, see litter box
  mice brought home by, 156
  moving and, 180
  neighborhood and, 32, 202
  spraying by, 117–119
  walking on leash, 158–161
infants, 193
infections, 175
injuries, 69, 102
insecticides, 202

Jacobson Organ, 20
Jai, 185
Java, 203
jumping onto chair or spot, teaching
    of, 137

Kali, 20, 35, 91, 93, 136, 146, 155
Kehei, 40–42, 43, 172, 190, 208
kennel, 30, 31

Kezar, 187–188, 189
Kimra, 12, 73, 82, 190–191
kitchen and dinnertime tips, 130–135
  see also feeding, food
kittens, 32, 34–35, 37–38, 97, 181
  conditioning behavior of, 37–38,
      56–58; see also conditioning of
      cat's behavior
  feeding of, 129
kitty litter, 200, 201
knocking things over, 33, 78, 79, 124

landlord, 32–33
leash, walking on, 158–161
"leave it" command, 88
"leave me alone" signals, 101–102
leukemia, 175, 197
licking, 18–20
life-style, 29–33
litter, 200, 201
litter box, 32, 114–116
  cleaning of, 116, 201
loner behavior, 94, 95
longevity, 74, 199
loud cats, 93

Mackenzie, Compton, 51
Marine World Africa USA, 24
markings and color, 16–17
marking territory, 21, 115, 117, 118,
    122
meditation, 107–108
meow, 21, 22, 33, 96
Merriam, Eve, 47
Mery, Fernand, 48, 91
mice, 156, 188
mirror cleaner, 202
Mivart, St. George Jackson, 60
Mohan, 12, 18, 19–20, 46, 70–71,
      96–97, 126, 131, 182, 203, 204
Mole, 26–27, 45, 149
moods, 102
mothballs, 120
mother figure, cat owner as, 159
  communication and, 97
  in conditioning kitten, 53–54, 55,
      58, 68, 133, 135
  gifts and, 156
  and introducing new cat, 183

outdoor cats and, 138, 144
  quiet time and, 94
mouth, 39, 102
movement, in play, 80–82
moving, 165, 178–180

naps, 91–94, 95, 96, 98
neighborhood, 32, 142, 202
neutering, 118, 143, 198–200
  advantages of, 199–200
  cat fights and, 32, 153, 154, 181, 200
  new pet and, 180–181
new situations, 177–194
  guests in house, 191–194
  moving, 165, 178–180
  new members of household, 191–194
  new pet, see pets, introducing cat to
  "no," 59, 61–62, 66, 68, 119, 133, 135
nose, 39, 102
nursing cats, feeding of, 129

outdoor cat, 113, 114, 141
  calling home, 138
  cold weather and, 150
  feeding of, 150; see also feeding, food
  see also indoor/outdoor cat
outdoors:
  introducing cat to, 143–147
  playtime and, 79
outside stimuli, aggressive behavior
    and, 69–71

panther, Florida, 206–207
pantry doors, 132, 134–135
paper bags, 85
Paradis de Moncrif, F. A., 165
partial vision, in play, 82
partner breathing, 108–111
patience, 60, 69, 105
paws, 103, 150
pedigree cats, 45
Penguin, 45, 113, 149
people, introducing cat to, 177–178,
    192
pets, introducing cat to, 180–191
  new cat, 182–185
  new cat to dog, 186–187
  new dog, 188
  smaller animals, 188–191

petting, 98, 99
  see also touch
physical checkup, during quiet time,
    102–103
Ping-Pong balls, 85
plants, 124–126
plastic wrap and bags, 120
play, 34
playtime, 12, 73–89, 135
  breakable items and, 78, 79, 124
  claws and, 69
  creativity in, 79–84
  etiquette in, 75–78
  fitness and well-being fostered by,
    88–89
  hunting skills and, 75, 88, 146, 157
  importance of, 73–75
  location of, 78–79
  sleep and, 94
  see also toys
poisonous plants, 125
poisons, 142, 150, 202
positive reinforcement, 62
  see also praise
possessiveness, 86–88
praise, 62, 66, 76, 78, 88, 123, 136
predatory instinct, 51, 131
  small pets and, 189, 190
pregnant cats, feeding of, 129
Project Tiger, 204
purring, 21, 96, 97, 107, 110

quality time, 13, 27, 111
  aggressive behavior and, 70
  and calling cat to come, 135
  with kitten, 56
  and responsibility of owning cat,
    29, 198
  see also playtime; quiet time
quiet time, 93, 94–95, 135
  conditioning cat to be cozy in, 98–101
  flea control and, 105–106
  grooming and, 104–105, 162
  "leave me alone" signals and, 101–
    102
  owner's benefits from, 94, 107, 111
  physical checkup in, 102–103
  talking to cat in, 96–97
  touch in, 97–98

rabbits, 188
rabies, 175
Rakon, 12, 35, 41, 78, 81, 83, 91–92,
    122, 135–136, 146, 155, 159,
    172, 182
relaxation, 95, 107, 108–109, 110, 111
  see also quiet time
repetition, in conditioning, 61
Repplier, Agnes, 38, 197
rewards, *see* treats and rewards
roar, 21, 22
rodents, as gifts from cat, 156
roommate, new, 192–193
Rousseau, Jean Jacques, 177
routine, change in, 129, 177–178
  see also new situations
rubbing, 21

Sampson, 12, 46, 86, 88, 182, 183, 185
scratching, 67–68
  of furniture, 121–122
  of people, 52, 64, 67–69, 75
scratching post, 121–123
Sennetty, 168–169
Serang, 35, 91, 146, 172, 190
shampoos, 202
shedding, 32, 104
shiny objects, 120
shots, *see* vaccinations
Siamese cats, 33
sight, sense of, 20–21, 80–82
similarities between tigers and cats,
    16–23
sleep, 91–94, 95, 96, 98
smell, sense of, 80, 83
sores, 69
sound, in play, 82–83
sounds, vocal, 21–22
spaying, 143, 181, 198–200
  advantages of, 199–200
  cat fights and, 153, 154
spouse, new, 192–193
spraying, 21, 115, 117–119, 122, 199
Stevenson, Adlai, 141
stimulation, 12, 74, 80, 88, 89
  with color, 80
  with movement, 80–82
  outdoors, 142–143, 145, 146
  with partial vision, 82

  with smell, 83
  with sound, 82–83
  see also playtime
stopping a game while it's still fun,
    83–84
Stowe, Harriet Beecher, 157
strangers, introducing cat to, 177–178,
    192
stray cats, adoption of, 36–37, 47–48
stress, in cat, 74, 88, 89
  new pet and, 180
  spraying and, 118
stress, in owner, 84
  quiet time and, 94, 107, 111
Sumo, 169

tail, 102
talking to your cat, 96–97
Tara, 12, 18, 33, 63, 185
tattoo identification, 143
teeth, 64, 65
  see also biting
territory, 144–146
  fights and, 153–154
  marking of, 21, 115, 117, 118, 122
thefts, pet, 143
therapy, animals as, 107
thinking like a cat, 60–61, 122, 124,
    169
tiger cubs, 34, 54–55, 57–58, 182–183
Tiger Island, 24–25, 205
tigers:
  endangerment of, 25, 48, 203–205
  as pets, 48–49
  physical similarities between cats
    and, 16–23
  world population of, 204
time, 60
  see also quality time
time-out, 61, 76, 78, 88
toilet lid, 120
tongue, 18–20
*Tonight Show (Starring Johnny Carson),
    The,* 187–188
touch, 54, 96–98
  health problems and, 102
  resistance toward, 98
toys, 77, 79, 82, 84–88, 202
  colors of, 80

toys *(cont.)*
    as gifts from cat, 156
    hands as, 75, 77
    possessiveness of, 86–88
    safe, 85
    and stopping a game while it's still
        fun, 83–84
trash and garbage, 121, 132, 150
travel box:
    for air travel, 173
    for car travel, 170
    conditioning cat to accept, 166–169
    in introducing new cat, 184–185
traveling, and leaving cat behind, 30–31
traveling, with cat, 129, 165–174
    by air, 173
    benefits of training cat for, 165
    in car, 170–172

moving, 165, 178–180
to veterinarian, 165, 166, 174
see also travel box
treats and rewards, 132, 136, 138–139,
    153
tuna cat food, 202
Turbo, 75
two cats, owning one vs., 34–36

urine, spraying of, 21, 115, 117–119,
    122, 199

vaccinations, 143, 174, 175
    new pet and, 180
veterinarian, 31, 39, 102, 103, 127,
    171, 175
    stray cats and, 47–48
    trips to, 165, 166, 174

vision, *see* sight, sense of
visualization, 110
vocalizations, 33

walking away, during playtime, 76–77
walking on leash, 158–161
water, 23, 41–42, 161, 162, 203
    baths, 18, 161–163
Weigall, Arthur, 135
weight, 127, 129
where to get a cat, 45–48
whiskers, 20
window cleaner, 202
windows, 120
wood surface scratches or stains,
    repair of, 202
words, learning of, 96
World Wildlife Fund, 204

ABOUT MARINE WORLD AFRICA USA

Tiger Island is just one of many innovative animal attractions found at Marine World Africa USA, a combination wildlife park and oceanarium, in Vallejo, California. It is the only park of its kind in the world, a home to wildlife of land, sea, and air and a unique celebration of mankind's relationship with the animal kingdom.

The deep bonds of mutual trust, respect, and affection showcased on Tiger Island are found throughout Marine World in many other close relationships of trainers with animals ranging from killer whales and dolphins to elephants and chimpanzees.

As a result, those who visit the park can experience inimitable firsthand encounters with animals, spectacular show performances, and an unforgettable introduction to the wonders of the animal world.

Marine World Africa USA is a nonprofit organization devoted to furthering people's understanding and appreciation of our world's wildlife. If you would like to find out more about this one-of-a-kind wildlife park, call (707) 644-4000 or write to Marine World Africa USA, Marine World Parkway, Vallejo, CA 94589.